Wind Energy: Theory and Applications

Wind Energy: Theory and Applications

Edited by
Jayce McCarthy

Larsen & Keller
www.larsen-keller.com

Wind Energy: Theory and Applications
Edited by Jayce McCarthy
ISBN: 978-1-63549-298-9 (Hardback)

© 2017 Larsen & Keller

⊟ Larsen & Keller

Published by Larsen and Keller Education,
5 Penn Plaza,
19th Floor,
New York, NY 10001, USA

Cataloging-in-Publication Data

Wind energy : theory and applications / edited by Jayce McCarthy.
   p. cm.
Includes bibliographical references and index.
ISBN 978-1-63549-298-9
   1. Wind power. 2. Wind power--Environmental aspects. 3. Wind power--Law and legislation.
4. Renewable energy sources. I. McCarthy, Jayce.
TJ820 .W56 2017
621.45--dc23

The publisher's policy is to use permanent paper from mills that operate a sustainable forestry policy. Furthermore, the publisher ensures that the text paper and cover boards used have met acceptable environmental accreditation standards.

Printed and bound in the United States of America.

For more information regarding Larsen and Keller Education and its products, please visit the publisher's website www.larsen-keller.com

Table of Contents

Preface

This book aims to shed light on some of the unexplored aspects of wind energy. It describes in detail the importance of wind power in the present scenario. Wind power refers to the science and technology of using wind turbines to generate electricity. It is a renewable source of energy and is an alternative for fossil fuels. This textbook is a compilation of chapters that discuss the most vital concepts in the field of wind power. It outlines the processes and applications of the subject in detail. Most of the themes introduced in this book cover fundamental techniques and application of wind energy. Coherent flow of topics, student-friendly language and extensive use of examples make the book an invaluable source of knowledge.

A short introduction to every chapter is written below to provide an overview of the content of the book:

Chapter 1 - Wind power is the power generated by wind with the help of wind turbine technology. It can be used as an alternative for fossil fuel. Wind power is preferred for it is renewable, clean and produces no greenhouse gas. The chapter on wind power offers an insightful focus, keeping in mind the complex subject matter; **Chapter 2** - The techniques used for the generation of wind powers are wind farms, offshore wind power, high-altitude wind power, windmills and windpumps. Wind farms consist of several individual wind turbines whereas the wind farms constructed on continental shelfs are known as offshore wind power. The aspects elucidated in the text are of vital importance, and provide a better understanding of wind power; **Chapter 3** - Wind turbine is a device that converts wind into electrical power. They are manufactured in a wide range; the range differs from vertical to horizontal axis types. Some of the aspects discussed within this chapter are wind turbine design, airborne wind turbine, vertical axis wind turbine, floating wind turbine and unconventional wind turbines. This chapter is an overview of the subject matter incorporating all the major aspects of wind turbines; **Chapter 4** - Wind power can be harnessed using wind turbines to power mechanical vehicles. These vehicles have been dubbed as 'land yachts'. An overview of crafts and vehicles running in the ocean and on land that operate on wind power is provided in this chapter; **Chapter 5** - The chapter concentrates on two basic laws related to wind power. Bert's law and wind profile power law; Bert's law directs the maximum power that can be extracted by wind. Wind profile power law is the law that forms a relation between the speed of wind at one height, and the same at another. This section serves as a source to understand the major laws related to wind power; **Chapter 6** - Wind power can be utilized in several ways in the near coming future. Some of the future prospects discussed in the following chapter are wind power forecasting, wind hybrid power systems and wind resource assessment. Wind power forecast provides an estimation of the expected production of one or more wind turbines. Wind power is emerging as a technology; the following chapter will not provide an overview, it will also delve into the topics related to it; **Chapter 7 -** Wind power generates the least global warming as per unit of electrical energy produced. Wind power consumes less land, generates less greenhouse gas and is also renewable. In comparison the effect wind power has to the effect is relatively minor to the effect fossil fuels have to the environment. This section explains to the reader the importance of wind power in contemporary times; **Chapter 8 -** The history of wind power can be traced by the evolution of devices that use this energy. Wind mills, sail ships are important landmarks in human technological history. This chapter provides an overview of the history of wind power and wind power related devices.

Finally, I would like to thank my fellow scholars who gave constructive feedback and my family members who supported me at every step.

Editor

Introduction to Wind Power

Wind power is the power generated by wind with the help of wind turbine technology. It can be used as an alternative for fossil fuel. Wind power is preferred for it is renewable, clean and produces no greenhouse gas. The chapter on wind power offers an insightful focus, keeping in mind the complex subject matter.

Wind power stations in Xinjiang, China

Wind power is the use of air flow through wind turbines to mechanically power generators for electricity. Wind power, as an alternative to burning fossil fuels, is plentiful, renewable, widely distributed, clean, produces no greenhouse gas emissions during operation, uses no water, and uses little land. The net effects on the environment are far less problematic than those of nonrenewable power sources.

Wind farms consist of many individual wind turbines which are connected to the electric power transmission network. Onshore wind is an inexpensive source of electricity, competitive with or in many places cheaper than coal or gas plants. Offshore wind is steadier and stronger than on land, and offshore farms have less visual impact, but construction and maintenance costs are considerably higher. Small onshore wind farms can feed some energy into the grid or provide electricity to isolated off-grid locations.

Wind power gives variable power which is very consistent from year to year but which has significant variation over shorter time scales. It is therefore used in conjunction with other electric power sources to give a reliable supply. As the proportion of wind power in a region increases, a need to upgrade the grid, and a lowered ability to supplant conventional production can occur. Power management techniques such as having excess capacity, geographically distributed turbines, dispatchable backing sources, sufficient hydroelectric power, exporting and importing power to neighboring areas, using vehicle-to-grid strategies or reducing demand when wind production is low, can in many cases overcome these problems. In addition, weather forecasting permits the electricity network to be readied for the predictable variations in production that occur.

As of 2015, Denmark generates 40% of its electricity from wind, and at least 83 other countries around the world are using wind power to supply their electricity grids. In 2014 global wind power capacity expanded 16% to 369,553 MW. Yearly wind energy production is also growing rapidly and has reached around 4% of worldwide electricity usage, 11.4% in the EU.

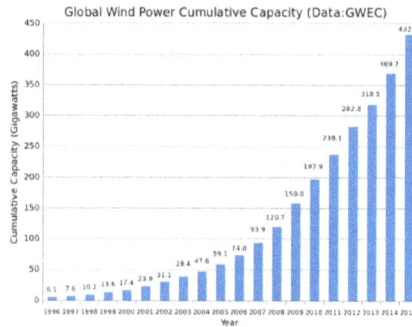

Global growth of installed capacity

History

Charles Brush's windmill of 1888, used for generating electricity.

Wind power has been used as long as humans have put sails into the wind. For more than two millennia wind-powered machines have ground grain and pumped water. Wind power was widely available and not confined to the banks of fast-flowing streams, or later, requiring sources of fuel. Wind-powered pumps drained the polders of the Netherlands, and in arid regions such as the American mid-west or the Australian out-back, wind pumps provided water for live stock and steam engines.

The first windmill used for the production of electricity was built in Scotland in July 1887 by Prof James Blyth of Anderson's College, Glasgow (the precursor of Strathclyde University). Blyth's 10 m high, cloth-sailed wind turbine was installed in the garden of his holiday cottage at Marykirk in Kincardineshire and was used to charge accumulators developed by the Frenchman Camille Alphonse Faure, to power the lighting in the cottage, thus making it the first house in the world to have its electricity supplied by wind power. Blyth offered the surplus electricity to the people of Marykirk for lighting the main street, however, they turned down the offer as they thought electricity was "the work of the devil." Although he later built a wind turbine to supply emergency power to the local Lunatic Asylum, Infirmary and Dispensary of Montrose the invention never really caught on as the technology was not considered to be economically viable.

Across the Atlantic, in Cleveland, Ohio a larger and heavily engineered machine was designed and constructed in the winter of 1887–1888 by Charles F. Brush, this was built by his engineering company at his home and operated from 1886 until 1900. The Brush wind turbine had a rotor 17 m (56 foot) in diameter and was mounted on an 18 m (60 foot) tower. Although large by today's standards, the machine was only rated at 12 kW. The connected dynamo was used either to charge a bank of batteries or to operate up to 100 incandescent light bulbs, three arc lamps, and various motors in Brush's laboratory.

With the development of electric power, wind power found new applications in lighting buildings remote from centrally-generated power. Throughout the 20th century parallel paths developed small wind stations suitable for farms or residences, and larger utility-scale wind generators that could be connected to electricity grids for remote use of power. Today wind powered generators operate in every size range between tiny stations for battery charging at isolated residences, up to near-gigawatt sized offshore wind farms that provide electricity to national electrical networks.

Wind Farms

Large onshore wind farms		
Wind farm	Current capacity (MW)	Country
Gansu Wind Farm	6,000	China
Muppandal wind farm	1,500	India
Alta (Oak Creek-Mojave)	1,320	United States
Jaisalmer Wind Park	1,064	India
Shepherds Flat Wind Farm	845	United States
Roscoe Wind Farm	782	United States
Horse Hollow Wind Energy Center	736	United States

Large onshore wind farms		
Wind farm	Current capacity (MW)	Country
Capricorn Ridge Wind Farm	662	United States
Fântânele-Cogealac Wind Farm	600	Romania
Fowler Ridge Wind Farm	600	United States
Whitelee Wind Farm	539	United Kingdom

A wind farm is a group of wind turbines in the same location used for production of electricity. A large wind farm may consist of several hundred individual wind turbines distributed over an extended area, but the land between the turbines may be used for agricultural or other purposes. For example, Gansu Wind Farm, the largest wind farm in the world, has several thousand turbines. A wind farm may also be located offshore.

Almost all large wind turbines have the same design — a horizontal axis wind turbine having an upwind rotor with three blades, attached to a nacelle on top of a tall tubular tower.

In a wind farm, individual turbines are interconnected with a medium voltage (often 34.5 kV), power collection system and communications network. In general, a distance of 7D (7 × Rotor Diameter of the Wind Turbine) is set between each turbine in a fully developed wind farm. At a substation, this medium-voltage electric current is increased in voltage with a transformer for connection to the high voltage electric power transmission system.

Generator Characteristics and Stability

Induction generators, which were often used for wind power projects in the 1980s and 1990s, require reactive power for excitation so substations used in wind-power collection systems include substantial capacitor banks for power factor correction. Different types of wind turbine generators behave differently during transmission grid disturbances, so extensive modelling of the dynamic electromechanical characteristics of a new wind farm is required by transmission system operators to ensure predictable stable behaviour during system faults. In particular, induction generators cannot support the system voltage during faults, unlike steam or hydro turbine-driven synchronous generators.

Today these generators aren't used any more in modern turbines. Instead today most turbines use variable speed generators combined with partial- or full-scale power converter between the turbine generator and the collector system, which generally have more desirable properties for grid interconnection and have Low voltage ride through-capabilities. Modern concepts use either doubly fed machines with partial-scale converters or squirrel-cage induction generators or synchronous generators (both permanently and electrically excited) with full scale converters.

Transmission systems operators will supply a wind farm developer with a grid code to specify the requirements for interconnection to the transmission grid. This will include power factor, constancy of frequency and dynamic behaviour of the wind farm turbines during a system fault.

Offshore Wind Power

The world's second full-scale floating wind turbine (and first to be installed without the use of heavy-lift vessels), WindFloat, operating at rated capacity (2 MW) approximately 5 km offshore of Póvoa de Varzim, Portugal

Offshore wind power refers to the construction of wind farms in large bodies of water to generate electricity. These installations can utilize the more frequent and powerful winds that are available in these locations and have less aesthetic impact on the landscape than land based projects. However, the construction and the maintenance costs are considerably higher.

Siemens and Vestas are the leading turbine suppliers for offshore wind power. DONG Energy, Vattenfall and E.ON are the leading offshore operators. As of October 2010, 3.16 GW of offshore wind power capacity was operational, mainly in Northern Europe. According to BTM Consult, more than 16 GW of additional capacity will be installed before the end of 2014 and the UK and Germany will become the two leading markets. Offshore wind power capacity is expected to reach a total of 75 GW worldwide by 2020, with significant contributions from China and the US.

In 2012, 1,662 turbines at 55 offshore wind farms in 10 European countries produced 18 TWh, enough to power almost five million households. As of August 2013 the London Array in the United Kingdom is the largest offshore wind farm in the world at 630 MW. This is followed by Gwynt y Môr (576 MW), also in the UK.

World's largest offshore wind farms				
Wind farm	Capacity (MW)	Country	Turbines and model	Commissioned
London Array	630	🇬🇧 United Kingdom	175 × Siemens SWT-3.6	2012

World's largest offshore wind farms				
Wind farm	Capacity (MW)	Country	Turbines and model	Commissioned
Gwynt y Môr	576	🇬🇧 United Kingdom	160 × Siemens SWT-3.6 107	2015
Greater Gabbard	504	🇬🇧 United Kingdom	140 × Siemens SWT-3.6	2012
Anholt	400	🇩🇰 Denmark	111 × Siemens SWT-3.6–120	2013
BARD Offshore 1	400	🇩🇪 Germany	80 BARD 5.0 turbines	2013

Collection and Transmission Network

In a wind farm, individual turbines are interconnected with a medium voltage (usually 34.5 kV) power collection system and communications network. At a substation, this medium-voltage electric current is increased in voltage with a transformer for connection to the high voltage electric power transmission system.

A transmission line is required to bring the generated power to (often remote) markets. For an off-shore station this may require a submarine cable. Construction of a new high-voltage line may be too costly for the wind resource alone, but wind sites may take advantage of lines installed for conventionally fueled generation.

One of the biggest current challenges to wind power grid integration in the United States is the necessity of developing new transmission lines to carry power from wind farms, usually in remote lowly populated states in the middle of the country due to availability of wind, to high load locations, usually on the coasts where population density is higher. The current transmission lines in remote locations were not designed for the transport of large amounts of energy. As transmission lines become longer the losses associated with power transmission increase, as modes of losses at lower lengths are exacerbated and new modes of losses are no longer negligible as the length is increased, making it harder to transport large loads over large distances. However, resistance from state and local governments makes it difficult to construct new transmission lines. Multi state power transmission projects are discouraged by states with cheap electricity rates for fear that exporting their cheap power will lead to increased rates. A 2005 energy law gave the Energy Department authority to approve transmission projects states refused to act on, but after an attempt to use this authority, the Senate declared the department was being overly aggressive in doing so. Another problem is that wind companies find out after the fact that the transmission capacity of a new farm is below the generation capacity, largely because federal utility rules to encourage renewable energy installation allow feeder lines to meet only minimum standards. These are important issues that need to be solved, as when the transmission capacity does not meet the generation capacity, wind farms are forced to produce below their full potential or stop running all together, in a process known as curtailment. While this leads to potential renewable generation left untapped, it prevents possible grid overload or risk to reliable service.

Wind Power Capacity and Production

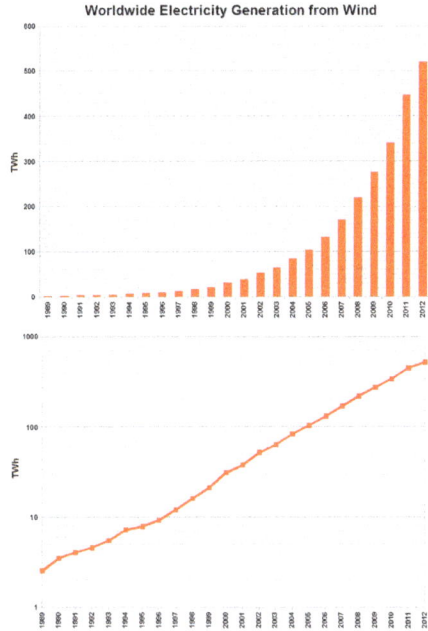

Worldwide wind generation up to 2012 (Source EIA, January 2015).

Worldwide there are now over two hundred thousand wind turbines operating, with a total nameplate capacity of 432,000 MW as of end 2015. The European Union alone passed some 100,000 MW nameplate capacity in September 2012, while the United States surpassed 75,000 MW in 2015 and China's grid connected capacity passed 145,000 MW in 2015.

World wind generation capacity more than quadrupled between 2000 and 2006, doubling about every three years. The United States pioneered wind farms and led the world in installed capacity in the 1980s and into the 1990s. In 1997 installed capacity in Germany surpassed the U.S. and led until once again overtaken by the U.S. in 2008. China has been rapidly expanding its wind installations in the late 2000s and passed the U.S. in 2010 to become the world leader. As of 2011, 83 countries around the world were using wind power on a commercial basis.

Wind power capacity has expanded rapidly to 336 GW in June 2014, and wind energy production was around 4% of total worldwide electricity usage, and growing rapidly. The actual amount of electricity that wind is able to generate is calculated by multiplying the nameplate capacity by the capacity factor, which varies according to equipment and location. Estimates of the capacity factors for wind installations are in the range of 35% to 44%.

Europe accounted for 48% of the world total wind power generation capacity in 2009. In 2010, Spain became Europe's leading producer of wind energy, achieving 42,976 GWh. Germany held the top spot in Europe in terms of installed capacity, with a total of 27,215 MW as of 31 December 2010. In 2015 wind power constituted 15.6% of all installed power generation capacity in the EU and it generates around 11.4% of its power.

Top windpower electricity producing countries in 2012 (TWh)		
Country	Windpower Production	% of World Total
United States	140.9	26.4
China	118.1	22.1
Spain	49.1	9.2
Germany	46.0	8.6
India	30.0	5.6
United Kingdom	19.6	3.7
France	14.9	2.8
Italy	13.4	2.5
Canada	11.8	2.2
Denmark	10.3	1.9
(rest of world)	80.2	15.0
World Total	534.3 TWh	100%
Source:Observ'ER – Electricity Production From Wind Sources [2012]		

Growth Trends

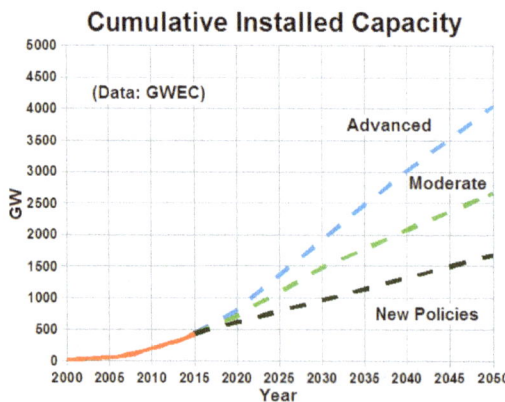

Worldwide installed wind power capacity forecast (Source: Global Wind Energy Council)

After setting new records in 2014, the wind power industry surprised many observers with another record breaking year in 2015, chalking up 22% annual market growth and passing the 60 GW mark for the first time in a single year; and this after having broken the 50 GW mark for the first time in 2014. In 2015, close to half of all new wind power was added outside of the traditional markets in Europe and North America. This was largely from new construction in China and India. Global Wind Energy Council (GWEC) figures show that 2015 recorded an increase of installed capacity of more than 63 GW, taking the total installed wind energy capacity to 432.9 GW, up from 74 GW in 2006. In terms of economic value, the wind energy sector has become one of the important players in the energy markets, with the total investments reaching US\$329bn (€296.6bn), an increase of 4% over 2014.

Although the wind power industry was affected by the global financial crisis in 2009 and 2010, GWEC predicts that the installed capacity of wind power will be 792.1 GW by the end of 2020 and 4,042 GW by end of 2050. The increased commissioning of wind power is being accompanied by record low prices for forthcoming renewable electricity. In some cases, wind onshore is already the cheapest electricity generation option and costs are continuing to decline. The contracted prices for wind onshore for the next few years are now as low as 30 USD/MWh.

In the EU in 2015, 44% of all new generating capacity was wind power; while in the same period net fossil fuel power capacity decreased.

Capacity Factor

Since wind speed is not constant, a wind farm's annual energy production is never as much as the sum of the generator nameplate ratings multiplied by the total hours in a year. The ratio of actual productivity in a year to this theoretical maximum is called the capacity factor. Typical capacity factors are 15–50%; values at the upper end of the range are achieved in favourable sites and are due to wind turbine design improvements.

Online data is available for some locations, and the capacity factor can be calculated from the yearly output. For example, the German nationwide average wind power capacity factor over all of 2012 was just under 17.5% (45867 GW·h/yr / (29.9 GW × 24 × 366) = 0.1746), and the capacity factor for Scottish wind farms averaged 24% between 2008 and 2010.

Unlike fueled generating plants, the capacity factor is affected by several parameters, including the variability of the wind at the site and the size of the generator relative to the turbine's swept area. A small generator would be cheaper and achieve a higher capacity factor but would produce less electricity (and thus less profit) in high winds. Conversely, a large generator would cost more but generate little extra power and, depending on the type, may stall out at low wind speed. Thus an optimum capacity factor of around 40–50% would be aimed for.

A 2008 study released by the U.S. Department of Energy noted that the capacity factor of new wind installations was increasing as the technology improves, and projected further improvements for future capacity factors. In 2010, the department estimated the capacity factor of new wind turbines in 2010 to be 45%. The annual average capacity factor for wind generation in the US has varied between 29.8% and 34.0% during the period 2010–2015.

Penetration

Country	Penetration
Denmark (2015)	42.1%
Portugal (2011)	19%
Spain (2011)	16%
Ireland (2012)	16%
United Kingdom (2014)	9.3%
Germany (2011)	8%
United States (2013)	4.5%

Wind energy penetration is the fraction of energy produced by wind compared with the total generation. There is no generally accepted maximum level of wind penetration. The limit for a particular grid will depend on the existing generating plants, pricing mechanisms, capacity for energy storage, demand management and other factors. An interconnected electricity grid will already include reserve generating and transmission capacity to allow for equipment failures. This reserve capacity can also serve to compensate for the varying power generation produced by wind stations. Studies have indicated that 20% of the total annual electrical energy consumption may be incorporated with minimal difficulty. These studies have been for locations with geographically dispersed wind farms, some degree of dispatchable energy or hydropower with storage capacity, demand management, and interconnected to a large grid area enabling the export of electricity when needed. Beyond the 20% level, there are few technical limits, but the economic implications become more significant. Electrical utilities continue to study the effects of large scale penetration of wind generation on system stability and economics.

A wind energy penetration figure can be specified for different durations of time, but is often quoted annually. To obtain 100% from wind annually requires substantial long term storage or substantial interconnection to other systems which may already have substantial storage. On a monthly, weekly, daily, or hourly basis—or less—wind might supply as much as or more than 100% of current use, with the rest stored or exported. Seasonal industry might then take advantage of high wind and low usage times such as at night when wind output can exceed normal demand. Such industry might include production of silicon, aluminum, steel, or of natural gas, and hydrogen, and using future long term storage to facilitate 100% energy from variable renewable energy. Homes can also be programmed to accept extra electricity on demand, for example by remotely turning up water heater thermostats.

In Australia, the state of South Australia generates around half of the nation's wind power capacity. By the end of 2011 wind power in South Australia, championed by Premier (and Climate Change Minister) Mike Rann, reached 26% of the State's electricity generation, edging out coal for the first time. At this stage South Australia, with only 7.2% of Australia's population, had 54% of Australia's installed capacity.

Variability

Windmills are typically installed in favourable windy locations. In the image, wind power generators in Spain, near an Osborne bull.

Electricity generated from wind power can be highly variable at several different timescales: hourly, daily, or seasonally. Annual variation also exists, but is not as significant. Because instantaneous electrical generation and consumption must remain in balance to maintain grid stability, this variability can present substantial challenges to incorporating large amounts of wind power into a grid system. Intermittency and the non-dispatchable nature of wind energy production can raise costs for regulation, incremental operating reserve, and (at high penetration levels) could require an increase in the already existing energy demand management, load shedding, storage solutions or system interconnection with HVDC cables.

Fluctuations in load and allowance for failure of large fossil-fuel generating units require reserve capacity that can also compensate for variability of wind generation.

Increase in system operation costs, Euros per MWh, for 10% & 20% wind share		
Country	10%	20%
Germany	2.5	3.2
Denmark	0.4	0.8
Finland	0.3	1.5
Norway	0.1	0.3
Sweden	0.3	0.7

Wind power is variable, and during low wind periods it must be replaced by other power sources. Transmission networks presently cope with outages of other generation plants and daily changes in electrical demand, but the variability of intermittent power sources such as wind power, are unlike those of conventional power generation plants which, when scheduled to be operating, may be able to deliver their nameplate capacity around 95% of the time.

Presently, grid systems with large wind penetration require a small increase in the frequency of usage of natural gas spinning reserve power plants to prevent a loss of electricity in the event that conditions are not favorable for power production from the wind. At lower wind power grid penetration, this is less of an issue.

GE has installed a prototype wind turbine with onboard battery similar to that of an electric car, equivalent of 1 minute of production. Despite the small capacity, it is enough to guarantee that power output complies with forecast for 15 minutes, as the battery is used to eliminate the difference rather than provide full output. The increased predictability can be used to take wind power penetration from 20 to 30 or 40 per cent. The battery cost can be retrieved by selling burst power on demand and reducing backup needs from gas plants.

A report on Denmark's wind power noted that their wind power network provided less than 1% of average demand on 54 days during the year 2002. Wind power advocates argue that these periods of low wind can be dealt with by simply restarting existing power stations that have been held in readiness, or interlinking with HVDC. Electrical grids with slow-responding thermal power plants and without ties to networks with hydroelectric generation may have to limit the use of wind power. According to a 2007 Stanford University study published in the *Journal of Applied Meteorology and Climatology*, interconnecting ten or more wind farms can allow an average of 33% of the total energy produced (i.e. about 8% of total nameplate capacity) to be used as reliable, baseload electric power which can be relied on to handle peak loads, as long as minimum criteria are met for wind speed and turbine height.

Conversely, on particularly windy days, even with penetration levels of 16%, wind power generation can surpass all other electricity sources in a country. In Spain, in the early hours of 16 April 2012 wind power production reached the highest percentage of electricity production till then, at 60.46% of the total demand. In Denmark, which had power market penetration of 30% in 2013, over 90 hours, wind power generated 100% of the country's power, peaking at 122% of the country's demand at 2 am on 28 October.

A 2006 International Energy Agency forum presented costs for managing intermittency as a function of wind-energy's share of total capacity for several countries, as shown in the table on the right. Three reports on the wind variability in the UK issued in 2009, generally agree that variability of wind needs to be taken into account, but it does not make the grid unmanageable. The additional costs, which are modest, can be quantified.

The combination of diversifying variable renewables by type and location, forecasting their variation, and integrating them with dispatchable renewables, flexible fueled generators, and demand response can create a power system that has the potential to meet power supply needs reliably. Integrating ever-higher levels of renewables is being successfully demonstrated in the real world:

In 2009, eight American and three European authorities, writing in the leading electrical engineers' professional journal, didn't find "a credible and firm technical limit to the amount of wind energy that can be accommodated by electricity grids". In fact, not one of more than 200 international studies, nor official studies for the eastern and western U.S. regions, nor the International Energy Agency, has found major costs or technical barriers to reliably integrating up to 30% variable renewable supplies into the grid, and in some studies much more. – *Reinventing Fire*

Solar power tends to be complementary to wind. On daily to weekly timescales, high pressure areas tend to bring clear skies and low surface winds, whereas low pressure areas tend to be windier and cloudier. On seasonal timescales, solar energy peaks in summer, whereas in many areas wind energy is lower in summer and higher in winter. Thus the intermittencies of wind and solar power tend to cancel each other somewhat. In 2007 the Institute for Solar Energy Supply Technology of the University of Kassel pilot-tested a combined power plant linking solar, wind, biogas and hydrostorage to provide load-following power around the clock and throughout the year, entirely from renewable sources.

Predictability

Wind power forecasting methods are used, but predictability of any particular wind farm is low for short-term operation. For any particular generator there is an 80% chance that wind output will change less than 10% in an hour and a 40% chance that it will change 10% or more in 5 hours.

However, studies by Graham Sinden (2009) suggest that, in practice, the variations in thousands of wind turbines, spread out over several different sites and wind regimes, are smoothed. As the distance between sites increases, the correlation between wind speeds measured at those sites, decreases.

Thus, while the output from a single turbine can vary greatly and rapidly as local wind speeds vary, as more turbines are connected over larger and larger areas the average power output becomes less variable and more predictable.

Wind power hardly ever suffers major technical failures, since failures of individual wind turbines have hardly any effect on overall power, so that the distributed wind power is reliable and predictable, whereas conventional generators, while far less variable, can suffer major unpredictable outages.

Energy Storage

Typically, conventional hydroelectricity complements wind power very well. When the wind is blowing strongly, nearby hydroelectric stations can temporarily hold back their water. When the wind drops they can, provided they have the generation capacity, rapidly increase production to compensate. This gives a very even overall power supply and virtually no loss of energy and uses no more water.

The Sir Adam Beck Generating Complex at Niagara Falls, Canada, includes a large pumped-storage hydroelectricity reservoir. During hours of low electrical demand excess electrical grid power is used to pump water up into the reservoir, which then provides an extra 174 MW of electricity during periods of peak demand.

Alternatively, where a suitable head of water is not available, pumped-storage hydroelectricity or other forms of grid energy storage such as compressed air energy storage and thermal energy storage can store energy developed by high-wind periods and release it when needed. The type of storage needed depends on the wind penetration level – low penetration requires daily storage, and high penetration requires both short and long term storage – as long as a month or more. Stored energy increases the economic value of wind energy since it can be shifted to displace higher cost generation during peak demand periods. The potential revenue from this arbitrage can offset the cost and losses of storage; the cost of storage may add 25% to the cost of any wind energy stored but it is not envisaged that this would apply to a large proportion of wind energy generated. For example, in the UK, the 1.7 GW Dinorwig pumped-storage plant evens out electrical demand peaks, and allows base-load suppliers to run their plants more efficiently. Although pumped-storage power systems are only about 75% efficient, and have high installation costs, their low running costs and ability to reduce the required electrical base-load can save both fuel and total electrical generation costs.

In particular geographic regions, peak wind speeds may not coincide with peak demand for electrical power. In the U.S. states of California and Texas, for example, hot days in summer may have low wind speed and high electrical demand due to the use of air conditioning. Some utilities subsidize the purchase of geothermal heat pumps by their customers, to reduce electricity demand during the summer months by making air conditioning up to 70% more efficient; widespread adoption of this technology would better match electricity demand to wind availability in areas with hot summers and low summer winds. A possible future option may be to interconnect widely dispersed geographic areas with an HVDC "super grid". In the U.S. it is estimated that to upgrade the transmission system to take in planned or potential renewables would cost at least $60 billion, while the society value of added windpower would be more than that cost.

Germany has an installed capacity of wind and solar that can exceed daily demand, and has been exporting peak power to neighboring countries, with exports which amounted to some 14.7 billion kilowatt hours in 2012. A more practical solution is the installation of thirty days storage capacity able to supply 80% of demand, which will become necessary when most of Europe's energy is obtained from wind power and solar power. Just as the EU requires member countries to maintain 90 days strategic reserves of oil it can be expected that countries will provide electricity storage, instead of expecting to use their neighbors for net metering.

Capacity Credit, Fuel Savings and Energy Payback

The capacity credit of wind is estimated by determining the capacity of conventional plants displaced by wind power, whilst maintaining the same degree of system security. According to the American Wind Energy Association, production of wind power in the United States in 2015 avoided consumption of 73 billion gallons of water and reduced CO_2 emissions by 132 million metric tons, while providing $7.3 billion in public health savings.

The energy needed to build a wind farm divided into the total output over its life, Energy Return on Energy Invested, of wind power varies but averages about 20–25. Thus, the energy payback time is typically around one year.

Economics

Wind turbines reached grid parity (the point at which the cost of wind power matches traditional sources) in some areas of Europe in the mid-2000s, and in the US around the same time. Falling prices continue to drive the levelized cost down and it has been suggested that it has reached general grid parity in Europe in 2010, and will reach the same point in the US around 2016 due to an expected reduction in capital costs of about 12%.

Electricity Cost and Trends

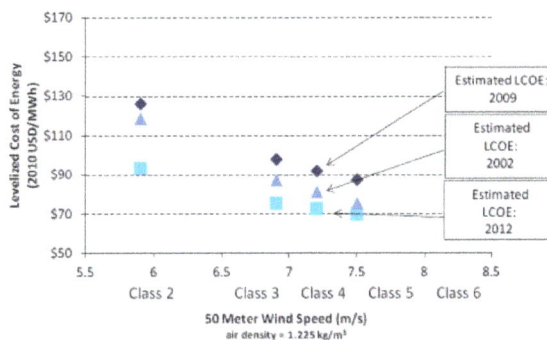

Estimated cost per MWh for wind power in Denmark

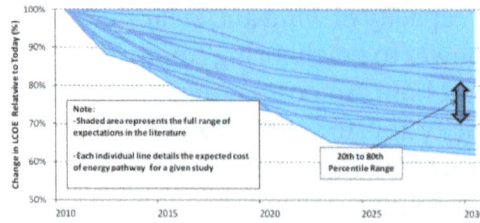

The National Renewable Energy Laboratory projects that the levelized cost of wind power in the U.S. will decline about 25% from 2012 to 2030.

A turbine blade convoy passing through Edenfield in the U.K. (2008). Even longer two-piece blades are now manufactured, and then assembled on-site to reduce difficulties in transportation.

Wind power is capital intensive, but has no fuel costs. The price of wind power is therefore much more stable than the volatile prices of fossil fuel sources. The marginal cost of wind energy once a station is constructed is usually less than 1-cent per kW·h.

However, the estimated average cost per unit of electricity must incorporate the cost of construction of the turbine and transmission facilities, borrowed funds, return to investors (including cost of risk), estimated annual production, and other components, averaged over the projected useful life of the equipment, which may be in excess of twenty years. Energy cost estimates are highly dependent on these assumptions so published cost figures can differ substantially. In 2004, wind energy cost a fifth of what it did in the 1980s, and some expected that downward trend to continue as larger multi-megawatt turbines were mass-produced. In 2012 capital costs for wind turbines were substantially lower than 2008–2010 but still above 2002 levels. A 2011 report from the American Wind Energy Association stated, "Wind's costs have dropped over the past two years, in the range of 5 to 6 cents per kilowatt-hour recently.... about 2 cents cheaper than coal-fired electricity, and more projects were financed through debt arrangements than tax equity structures last year.... winning more mainstream acceptance from Wall Street's banks.... Equipment makers can also deliver products in the same year that they are ordered instead of waiting up to three years as was the case in previous cycles.... 5,600 MW of new installed capacity is under construction in the United States, more than double the number at this point in 2010. Thirty-five percent of all new power generation built in the United States since 2005 has come from wind, more than new gas and coal plants combined, as power providers are increasingly enticed to wind as a convenient hedge against unpredictable commodity price moves."

A British Wind Energy Association report gives an average generation cost of onshore wind power of around 3.2 pence (between US 5 and 6 cents) per kW·h (2005). Cost per unit of energy produced was estimated in 2006 to be 5 to 6 percent above the cost of new generating capacity in the US for coal and natural gas: wind cost was estimated at $55.80 per MW·h, coal at $53.10/MW·h and natural gas at $52.50. Similar comparative results with natural gas were obtained in a governmental study in the UK in 2011. In 2011 power from wind turbines could be already cheaper than fossil or nuclear plants; it is also expected that wind power will be the cheapest form of energy generation in the future. The presence of wind energy, even when subsidised, can reduce costs for consumers (€5 billion/yr in Germany) by reducing the marginal price, by minimising the use of expensive peaking power plants.

An 2012 EU study shows base cost of onshore wind power similar to coal, when subsidies and externalities are disregarded. Wind power has some of the lowest external costs.

In February 2013 Bloomberg New Energy Finance (BNEF) reported that the cost of generating electricity from new wind farms is cheaper than new coal or new baseload gas plants. When including the current Australian federal government carbon pricing scheme their modeling gives costs (in Australian dollars) of $80/MWh for new wind farms, $143/MWh for new coal plants and $116/MWh for new baseload gas plants. The modeling also shows that "even without a carbon price (the most efficient way to reduce economy-wide emissions) wind energy is 14% cheaper than new coal and 18% cheaper than new gas." Part of the higher costs for new coal plants is due to high financial lending costs because of "the reputational damage of emissions-intensive investments". The expense of gas fired plants is partly due to "export market" effects on local prices. Costs of production from coal fired plants built in "the 1970s and 1980s" are cheaper than renewable energy sources because of depreciation. In 2015 BNEF calculated LCOE prices per MWh energy in new powerplants (excluding carbon costs) : $85 for onshore wind ($175 for offshore), $66–75 for coal in the Americas ($82–105 in Europe), gas $80–100. A 2014 study showed unsubsidized LCOE costs between $37–81, depending on region. A 2014 US DOE report showed that in some cases power purchase agreement prices for wind power had dropped to record lows of $23.5/MWh.

The cost has reduced as wind turbine technology has improved. There are now longer and lighter wind turbine blades, improvements in turbine performance and increased power generation efficiency. Also, wind project capital and maintenance costs have continued to decline. For example, the wind industry in the USA in early 2014 were able to produce more power at lower cost by using taller wind turbines with longer blades, capturing the faster winds at higher elevations. This has opened up new opportunities and in Indiana, Michigan, and Ohio, the price of power from wind turbines built 300 feet to 400 feet above the ground can now compete with conventional fossil fuels like coal. Prices have fallen to about 4 cents per kilowatt-hour in some cases and utilities have been increasing the amount of wind energy in their portfolio, saying it is their cheapest option.

A number of initiatives are working to reduce costs of electricity from offshore wind. One example is the Carbon Trust Offshore Wind Accelerator, a joint industry project, involving nine offshore wind developers, which aims to reduce the cost of offshore wind by 10% by 2015. It has been suggested that innovation at scale could deliver 25% cost reduction in offshore wind by 2020. Henrik Stiesdal, former Chief Technical Officer at Siemens Wind Power, has stated that by 2025 energy from offshore wind will be one of the cheapest, scalable solutions in the UK, compared to other renewables and fossil fuel energy sources, if the true cost to society is factored into the cost of energy equation. He calculates the cost at that time to be 43 EUR/MWh for onshore, and 72 EUR/MWh for offshore wind.

Incentives and Community Benefits

U.S. landowners typically receive $3,000–$5,000 annual rental income per wind turbine, while farmers continue to grow crops or graze cattle up to the foot of the turbines. Shown: the Brazos Wind Farm, Texas.

Some of the 6,000 turbines in California's Altamont Pass Wind Farm aided by tax incentives during the 1980s.

The U.S. wind industry generates tens of thousands of jobs and billions of dollars of economic activity. Wind projects provide local taxes, or payments in lieu of taxes and strengthen the economy of rural communities by providing income to farmers with wind turbines on their land. Wind energy in many jurisdictions receives financial or other support to encourage its development. Wind energy benefits from subsidies in many jurisdictions, either to increase its attractiveness, or to compensate for subsidies received by other forms of production which have significant negative externalities.

In the US, wind power receives a production tax credit (PTC) of 1.5¢/kWh in 1993 dollars for each kW·h produced, for the first ten years; at 2.2 cents per kW·h in 2012, the credit was renewed on 2 January 2012, to include construction begun in 2013. A 30% tax credit can be applied instead of receiving the PTC. Another tax benefit is accelerated depreciation. Many American states also provide incentives, such as exemption from property tax, mandated purchases, and additional markets for "green credits". The Energy Improvement and Extension Act of 2008 contains extensions of credits for wind, including microturbines. Countries such as Canada and Germany also provide incentives for wind turbine construction, such as tax credits or minimum purchase prices for wind generation, with assured grid access (sometimes referred to as feed-in tariffs). These feed-in tariffs are typically set well above average electricity prices. In December 2013 U.S. Senator Lamar Alexander and other Republican senators argued that the "wind energy production tax credit should be allowed to expire at the end of 2013" and it expired 1 January 2014 for new installations.

Secondary market forces also provide incentives for businesses to use wind-generated power, even if there is a premium price for the electricity. For example, socially responsible manufacturers pay utility companies a premium that goes to subsidize and build new wind power infrastructure. Companies use wind-generated power, and in return they can claim that they are undertaking strong "green" efforts. In the US the organization Green-e monitors business compliance with these renewable energy credits.

Small-Scale Wind Power

A small Quietrevolution QR5 Gorlov type vertical axis wind turbine on the roof of Colston Hall in Bristol, England. Measuring 3 m in diameter and 5 m high, it has a nameplate rating of 6.5 kW.

Small-scale wind power is the name given to wind generation systems with the capacity to produce up to 50 kW of electrical power. Isolated communities, that may otherwise rely on diesel generators, may use wind turbines as an alternative. Individuals may purchase these systems to reduce or eliminate their dependence on grid electricity for economic reasons, or to reduce their carbon footprint. Wind turbines have been used for household electricity generation in conjunction with battery storage over many decades in remote areas.

Recent examples of small-scale wind power projects in an urban setting can be found in New York City, where, since 2009, a number of building projects have capped their roofs with Gorlov-type helical wind turbines. Although the energy they generate is small compared to the buildings' overall consumption, they help to reinforce the building's 'green' credentials in ways that "showing people your high-tech boiler" can not, with some of the projects also receiving the direct support of the New York State Energy Research and Development Authority.

Grid-connected domestic wind turbines may use grid energy storage, thus replacing purchased electricity with locally produced power when available. The surplus power produced by domestic microgenerators can, in some jurisdictions, be fed into the network and sold to the utility company, producing a retail credit for the microgenerators' owners to offset their energy costs.

Off-grid system users can either adapt to intermittent power or use batteries, photovoltaic or diesel systems to supplement the wind turbine. Equipment such as parking meters, traffic warning signs, street lighting, or wireless Internet gateways may be powered by a small wind turbine, possibly combined with a photovoltaic system, that charges a small battery replacing the need for a connection to the power grid.

A Carbon Trust study into the potential of small-scale wind energy in the UK, published in 2010, found that small wind turbines could provide up to 1.5 terawatt hours (TW·h) per year of electricity (0.4% of total UK electricity consumption), saving 0.6 million tonnes of carbon dioxide (Mt CO_2) emission savings. This is based on the assumption that 10% of households would install turbines at costs competitive with grid electricity, around 12 pence (US 19 cents) a kW·h. A report prepared for the UK's government-sponsored Energy Saving Trust in 2006, found that home power generators of various kinds could provide 30 to 40% of the country's electricity needs by 2050.

Distributed generation from renewable resources is increasing as a consequence of the increased awareness of climate change. The electronic interfaces required to connect renewable generation units with the utility system can include additional functions, such as the active filtering to enhance the power quality.

Environmental Effects

The environmental impact of wind power when compared to the environmental impacts of fossil fuels, is relatively minor. According to the IPCC, in assessments of the life-cycle global warming potential of energy sources, wind turbines have a median value of between 12 and 11 (gCO_{2eq}/kWh) depending on whether off- or onshore turbines are being assessed. Compared with other low carbon power sources, wind turbines have some of the lowest global warming potential per unit of electrical energy generated.

Livestock grazing near a wind turbine.

While a wind farm may cover a large area of land, many land uses such as agriculture are compatible with it, as only small areas of turbine foundations and infrastructure are made unavailable for use.

There are reports of bird and bat mortality at wind turbines as there are around other artificial structures. The scale of the ecological impact may or may not be significant, depending on specific circumstances. Prevention and mitigation of wildlife fatalities, and protection of peat bogs, affect the siting and operation of wind turbines.

Wind turbines generate some noise. At a residential distance of 300 metres (980 ft) this may be around 45 dB, which is slightly louder than a refrigerator. At 1.5 km (1 mi) distance they become inaudible. There are anecdotal reports of negative health effects from noise on people who live very close to wind turbines. Peer-reviewed research has generally not supported these claims.

The United States Air Force and Navy have expressed concern that siting large windmills near bases "will negatively impact radar to the point that air traffic controllers will lose the location of aircraft."

Aesthetic aspects of wind turbines and resulting changes of the visual landscape are significant. Conflicts arise especially in scenic and heritage protected landscapes.

Politics

Central Government

Nuclear power and fossil fuels are subsidized by many governments, and wind power and other forms of renewable energy are also often subsidized. For example, a 2009 study by the Environmental Law Institute assessed the size and structure of U.S. energy subsidies over the 2002–2008 period. The study estimated that subsidies to fossil-fuel based sources amounted to approximately $72 billion over this period and subsidies to renewable fuel sources totalled $29 billion. In the United States, the federal government has paid US$74 billion for energy subsidies to support R&D for nuclear power ($50 billion) and fossil fuels ($24 billion) from 1973 to 2003. During this same time frame, renewable energy technologies and energy efficiency received a total of US$26 billion. It has been suggested that a subsidy shift would help to level the playing field and support growing energy sectors, namely solar power, wind power, and biofuels. History shows that no energy sector was developed without subsidies.

Part of the Seto Hill Windfarm in Japan.

According to the International Energy Agency (IEA) (2011), energy subsidies artificially lower the price of energy paid by consumers, raise the price received by producers or lower the cost of production. "Fossil fuels subsidies costs generally outweigh the benefits. Subsidies to renewables and low-carbon energy technologies can bring long-term economic and environmental benefits". In November 2011, an IEA report entitled *Deploying Renewables 2011* said "subsidies in green energy technologies that were not yet competitive are justified in order to give an incentive to investing into technologies with clear environmental and energy security benefits". The IEA's report disagreed with claims that renewable energy technologies are only viable through costly subsidies and not able to produce energy reliably to meet demand.

In the U.S., the wind power industry has recently increased its lobbying efforts considerably, spending about $5 million in 2009 after years of relative obscurity in Washington. By comparison, the U.S. nuclear industry alone spent over $650 million on its lobbying efforts and campaign contributions during a single ten-year period ending in 2008.

Following the 2011 Japanese nuclear accidents, Germany's federal government is working on a new plan for increasing energy efficiency and renewable energy commercialization, with a particular focus on offshore wind farms. Under the plan, large wind turbines will be erected far away from the coastlines, where the wind blows more consistently than it does on land, and where the enormous turbines won't bother the inhabitants. The plan aims to decrease Germany's dependence on energy derived from coal and nuclear power plants.

Public Opinion

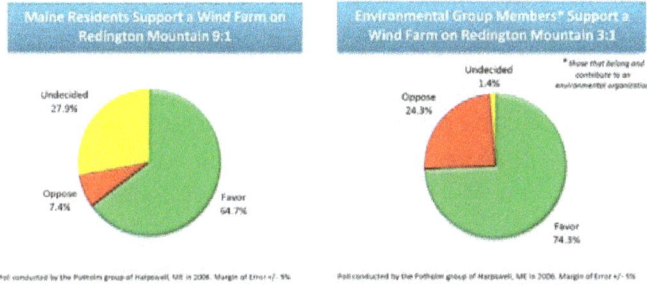

Environmental group members are both more in favor of wind power (74%) as well as more opposed (24%). Few are undecided.

Surveys of public attitudes across Europe and in many other countries show strong public support for wind power. About 80% of EU citizens support wind power. In Germany, where wind power has gained very high social acceptance, hundreds of thousands of people have invested in citizens' wind farms across the country and thousands of small and medium-sized enterprises are running successful businesses in a new sector that in 2008 employed 90,000 people and generated 8% of Germany's electricity.

Although wind power is a popular form of energy generation, the construction of wind farms is not universally welcomed, often for aesthetic reasons.

In Spain, with some exceptions, there has been little opposition to the installation of inland wind parks. However, the projects to build offshore parks have been more controversial. In particular, the proposal of building the biggest offshore wind power production facility in the world in southwestern Spain in the coast of Cádiz, on the spot of the 1805 Battle of Trafalgar has been met with strong opposition who fear for tourism and fisheries in the area, and because the area is a war grave.

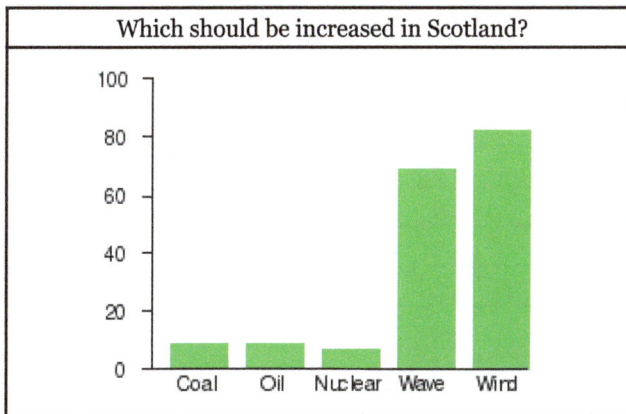

In a survey conducted by Angus Reid Strategies in October 2007, 89 per cent of respondents said that using renewable energy sources like wind or solar power was positive for Canada, because these sources were better for the environment. Only 4 per cent considered using renewable sources as negative since they can be unreliable and expensive. According to a Saint Consulting survey in April 2007, wind power was the alternative energy source most likely to gain public support for future development in Canada, with only 16% opposed to this type of energy. By contrast, 3 out of 4 Canadians opposed nuclear power developments.

A 2003 survey of residents living around Scotland's 10 existing wind farms found high levels of community acceptance and strong support for wind power, with much support from those who lived closest to the wind farms. The results of this survey support those of an earlier Scottish Executive survey 'Public attitudes to the Environment in Scotland 2002', which found that the Scottish public would prefer the majority of their electricity to come from renewables, and which rated wind power as the cleanest source of renewable energy. A survey conducted in 2005 showed that 74% of people in Scotland agree that wind farms are necessary to meet current and future energy needs. When people were asked the same question in a Scottish renewables study conducted in 2010, 78% agreed. The increase is significant as there were twice as many wind farms in 2010 as there were in 2005. The 2010 survey also showed that 52% disagreed with the statement that wind farms are "ugly and a blot on the landscape". 59% agreed that wind farms were necessary and that how they looked was unimportant. Regarding tourism, query responders consider power pylons, cell phone towers, quarries and plantations more negatively than wind farms. Scotland is planning to obtain 100% of electricity from renewable sources by 2020.

In other cases there is direct community ownership of wind farm projects. The hundreds of thousands of people who have become involved in Germany's small and medium-sized wind farms demonstrate such support there.

This 2010 Harris Poll reflects the strong support for wind power in Germany, other European countries, and the U.S.

Opinion on increase in number of wind farms, 2010 Harris Poll						
	U.S.	Great Britain	France	Italy	Spain	Germany
	%	%	%	%	%	%
Strongly oppose	3	6	6	2	2	4
Oppose more than favour	9	12	16	11	9	14
Favour more than oppose	37	44	44	38	37	42
Strongly favour	50	38	33	49	53	40

Community

Many wind power companies work with local communities to reduce environmental and other concerns associated with particular wind farms. In other cases there is direct community ownership of wind farm projects. Appropriate government consultation, planning and approval procedures also help to minimize environmental risks. Some may still object to wind farms but, according to The Australia Institute, their concerns should be weighed against the need to address the threats posed by climate change and the opinions of the broader community.

Wind turbines such as these, in Cumbria, England, have been opposed for a number of reasons, including aesthetics, by some sectors of the population.

In America, wind projects are reported to boost local tax bases, helping to pay for schools, roads and hospitals. Wind projects also revitalize the economy of rural communities by providing steady income to farmers and other landowners.

In the UK, both the National Trust and the Campaign to Protect Rural England have expressed concerns about the effects on the rural landscape caused by inappropriately sited wind turbines and wind farms.

A panoramic view of the United Kingdom's Whitelee Wind Farm with Lochgoin Reservoir in the foreground.

Some wind farms have become tourist attractions. The Whitelee Wind Farm Visitor Centre has an exhibition room, a learning hub, a café with a viewing deck and also a shop. It is run by the Glasgow Science Centre.

In Denmark, a loss-of-value scheme gives people the right to claim compensation for loss of value of their property if it is caused by proximity to a wind turbine. The loss must be at least 1% of the property's value.

Despite this general support for the concept of wind power in the public at large, local opposition often exists and has delayed or aborted a number of projects. For example, there are concerns that some installations can negatively affect TV and radio reception and Doppler weather radar, as well as produce excessive sound and vibration levels leading to a decrease in property values. Potential broadcast-reception solutions include predictive interference modeling as a component of site selection. A study of 50,000 home sales near wind turbines found no statistical evidence that prices were affected.

While aesthetic issues are subjective and some find wind farms pleasant and optimistic, or symbols of energy independence and local prosperity, protest groups are often formed to attempt to block new wind power sites for various reasons.

This type of opposition is often described as NIMBYism, but research carried out in 2009 found that there is little evidence to support the belief that residents only object to renewable power facilities such as wind turbines as a result of a "Not in my Back Yard" attitude.

Turbine Design

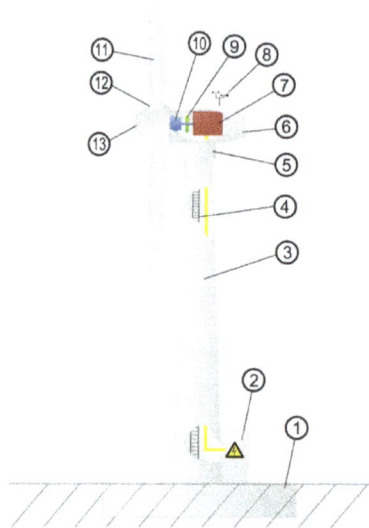

Typical wind turbine components : 1-Foundation, 2-Connection to the electric grid, 3-Tower, 4-Access ladder, 5-Wind orientation control (Yaw control), 6-Nacelle, 7-Generator, 8-Anemometer, 9-Electric or Mechanical Brake, 10-Gearbox, 11-Rotor blade, 12-Blade pitch control, 13-Rotor hub.

Typical components of a wind turbine (gearbox, rotor shaft and brake assembly) being lifted into position

Wind turbines are devices that convert the wind's kinetic energy into electrical power. The result of over a millennium of windmill development and modern engineering, today's wind turbines are manufactured in a wide range of horizontal axis and vertical axis types. The smallest turbines are used for applications such as battery charging for auxiliary power. Slightly larger turbines can be used for making small contributions to a domestic power supply while selling unused power back to the utility supplier via the electrical grid. Arrays of large turbines, known as wind farms, have become an increasingly important source of renewable energy and are used in many countries as part of a strategy to reduce their reliance on fossil fuels.

Wind turbine design is the process of defining the form and specifications of a wind turbine to extract energy from the wind. A wind turbine installation consists of the necessary systems needed to capture the wind's energy, point the turbine into the wind, convert mechanical rotation into electrical power, and other systems to start, stop, and control the turbine.

In 1919 the German physicist Albert Betz showed that for a hypothetical ideal wind-energy extraction machine, the fundamental laws of conservation of mass and energy allowed no more than 16/27 (59.3%) of the kinetic energy of the wind to be captured. This Betz limit can be approached in modern turbine designs, which may reach 70 to 80% of the theoretical Betz limit.

The aerodynamics of a wind turbine are not straightforward. The air flow at the blades is not the same as the airflow far away from the turbine. The very nature of the way in which energy is extracted from the air also causes air to be deflected by the turbine. In addition the aerodynamics of a wind turbine at the rotor surface exhibit phenomena that are rarely seen in other aerodynamic fields. The shape and dimensions of the blades of the wind turbine are determined by the aerodynamic performance required to efficiently extract energy from the wind, and by the strength required to resist the forces on the blade.

In addition to the aerodynamic design of the blades, the design of a complete wind power system must also address the design of the installation's rotor hub, nacelle, tower structure, generator, controls, and foundation. Further design factors must also be considered when integrating wind turbines into electrical power grids.

Wind Energy

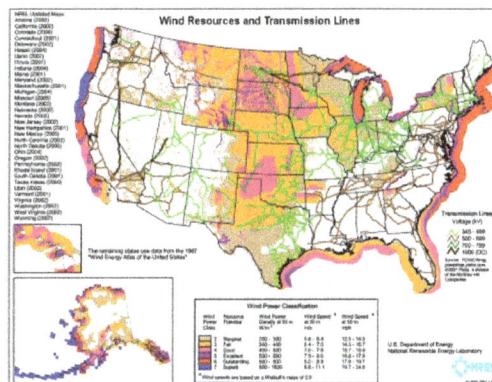

Map of available wind power for the United States. Color codes indicate wind power density class.

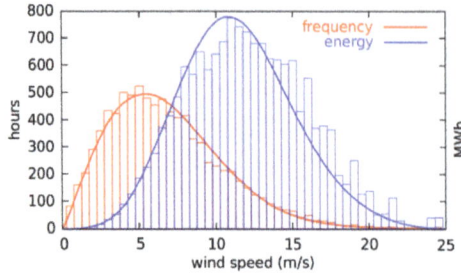

Distribution of wind speed (red) and energy (blue) for all of 2002 at the Lee Ranch facility in Colorado. The histogram shows measured data, while the curve is the Rayleigh model distribution for the same average wind speed.

Wind energy is the kinetic energy of air in motion, also called wind. Total wind energy flowing through an imaginary surface with area A during the time t is:

$$E = \frac{1}{2}mv^2 = \frac{1}{2}(Avt\rho)v^2 = \frac{1}{2}At\rho v^3,$$

where ρ is the density of air; v is the wind speed; Avt is the volume of air passing through A (which is considered perpendicular to the direction of the wind); $Avt\rho$ is therefore the mass m passing through "A". Note that $\frac{1}{2}\rho v^2$ is the kinetic energy of the moving air per unit volume.

Power is energy per unit time, so the wind power incident on A (e.g. equal to the rotor area of a wind turbine) is:

$$P = \frac{E}{t} = \frac{1}{2}A\rho v^3$$

Wind power in an open air stream is thus *proportional* to the *third power* of the wind speed; the available power increases eightfold when the wind speed doubles. Wind turbines for grid electricity therefore need to be especially efficient at greater wind speeds.

Wind is the movement of air across the surface of the Earth, affected by areas of high pressure and of low pressure. The global wind kinetic energy averaged approximately 1.50 MJ/m² over the period from 1979 to 2010, 1.31 MJ/m² in the Northern Hemisphere with 1.70 MJ/m² in the Southern Hemisphere. The atmosphere acts as a thermal engine, absorbing heat at higher temperatures, releasing heat at lower temperatures. The process is responsible for production of wind kinetic energy at a rate of 2.46 W/m² sustaining thus the circulation of the atmosphere against frictional dissipation. A global 1 km² map of wind resources is housed at http://irena.masdar.ac.ae/?map=103 , based on calculations by the Technical University of Denmark.

The total amount of economically extractable power available from the wind is considerably more than present human power use from all sources. Axel Kleidon of the Max Planck Institute in Germany, carried out a "top down" calculation on how much wind energy there is, starting with the incoming solar radiation that drives the winds by creating temperature differences in the atmosphere. He concluded that somewhere between 18 TW and 68 TW could be extracted.

Cristina Archer and Mark Z. Jacobson presented a "bottom-up" estimate, which unlike Kleidon's are based on actual measurements of wind speeds, and found that there is 1700 TW of wind power at an altitude of 100 metres over land and sea. Of this, "between 72 and 170 TW could be extracted in a practical and cost-competitive manner". They later estimated 80 TW. However research at Harvard University estimates 1 Watt/m² on average and 2–10 MW/km² capacity for large scale wind farms, suggesting that these estimates of total global wind resources are too high by a factor of about 4.

The strength of wind varies, and an average value for a given location does not alone indicate the amount of energy a wind turbine could produce there.

To assess prospective wind power sites a probability distribution function is often fit to the observed wind speed data. Different locations will have different wind speed distributions. The Weibull model closely mirrors the actual distribution of hourly/ten-minute wind speeds at many locations. The Weibull factor is often close to 2 and therefore a Rayleigh distribution can be used as a less accurate, but simpler model.

References

- History of Wind Energy in Cutler J. Cleveland,(ed) Encyclopedia of Energy Vol.6, Elsevier, ISBN 978-1-60119-433-6, 2007, pp. 421–422

- Gipe, Paul (1995). Wind Energy Comes of Age. John Wiley & Sons. pp. 376–. ISBN 978-0-471-10924-2.

- Wind Turbines in Denmark (PDF). section 6.8, p. 22, Danish Energy Agency. November 2009. ISBN 978-87-7844-821-7.

- Denmark breaks its own world record in wind energy. Euractiv.com (15 January 2016). Retrieved on 2016-07-20.

- New record-breaking year for Danish wind power. Energinet.dk (15 January 2016). Retrieved on 2016-07-20.

- Started in August 2001, the Jaisalmer based facility crossed 1,000 MW capacity to achieve this milestone. Business-standard.com (11 May 2012). Retrieved on 20 July 2016.

- CEZ Group: The Largest Wind Farm in Europe Goes Into Trial Operation. Cez.cz. Retrieved on 20 July 2016.

- Inadequate transmission lines keeping some Maine wind power off the grid – The Portland Press Herald / Maine Sunday Telegram. Pressherald.com (4 August 2013). Retrieved on 2016-07-20.

- "China now the world leader in wind power production". The Globe and Mail. 11 February 2016. Retrieved 28 February 2016.

- "Global Wind Report 2014 – Annual Market Update" (PDF). report. GWEC. 22 April 2016. Retrieved 23 May 2016.

- $7.3 billion in public health savings seen in 2015 from wind energy cutting air pollution. Awea.

org (29 March 2016). Retrieved on 2016-07-20.

- "Global wind expert says offshore wind will be one of the cheapest UK energy sources by 2025". The Carbon Trust. 23 September 2014. Retrieved 20 January 2015.

- Stiesdal, Henrik. "Den fremtidige pris på vindkraft" Ingeniøren, 13 September 2015. The future price of wind power

- Walwyn, David Richard; Brent, Alan Colin (2015). "Renewable energy gathers steam in South Africa". Renewable and Sustainable Energy Reviews. 41: 390. doi:10.1016/j.rser.2014.08.049.

- "World's second largest offshore wind farm opens off coast of Wales". Wales Online. Retrieved 18 June 2015.

- "International Energy Statistics". U.S. Energy Information Administration (EIA). Retrieved 26 January 2015.

- Electricity needs of more than a quarter of UK homes powered by wind in 2014. RenewableUK News (5 January 2015)

- "A New Era for Wind Power in the United States" page XIV. United States Department of Energy, 2013. Retrieved: March 2015.

- "Red Eléctrica de España | Wind produces more than 60% of the electricity consumed in Spain during the early hours of this morning". www.ree.es. Retrieved 27 July 2015.

- Professor Simon Chapman. "Summary of main conclusions reached in 25 reviews of the research literature on wind farms and health" Sydney University School of Public Health, April 2015

- Cramer, Glenn (30 October 2009). "Town Councilor regrets High Sheldon Wind Farm (Sheldon, NY)". Retrieved 4 September 2015.

- Broadcast Wind, LLC. "Solutions for the Broadcasting and Wind Energy Industries". Retrieved 4 September 2015.

- "IMPACT OF WIND FARMS ON RADIOCOMMUNICATION SERVICES". TSR (grupo Tratamiento de Señal y Radiocomunicaciones de la UPV/EHU). Retrieved 4 September 2015.

- Seeing the Wind with the New Global Wind Atlas International Renewable Energy Agency, 21 October 2015.

2

Wind Power Generation Techniques

The techniques used for the generation of wind powers are wind farms, offshore wind power, high-altitude wind power, windmills and windpumps. Wind farms consist of several individual wind turbines whereas the wind farms constructed on continental shelfs are known as offshore wind power. The aspects elucidated in the text are of vital importance, and provide a better understanding of wind power.

Wind Farm

The Gansu Wind Farm in China is the largest wind farm in the world, with a target capacity of 20,000 MW by 2020.

A wind farm or wind park is a group of wind turbines in the same location used to produce electricity. A large wind farm may consist of several hundred individual wind turbines and cover an extended area of hundreds of square miles, but the land between the turbines may be used for agricultural or other purposes. A wind farm can also be located offshore.

Many of the largest operational onshore wind farms are located in Germany, China and the United States. For example, the largest wind farm in the world, Gansu Wind Farm in China has a capacity of over 6,000 MW of power in 2012 with a goal of 20,000 MW by 2020. The Alta Wind Energy Center in California, United States is the largest on-shore wind farm outside of China, with a capacity of 1,020 MW. As of April 2013, the 630 MW London Array in the UK is the largest offshore wind farm in the world, followed by the 504 MW Greater Gabbard wind farm in the UK.

There are many large wind farms under construction and these include Sinus Holding Wind Farm (700 MW), Lincs Wind Farm (270 MW), Lower Snake River Wind Project (343 MW), Macarthur Wind Farm (420 MW).

The Shepherds Flat Wind Farm is an 845 MW wind farm in the U.S. state of Oregon.

Design

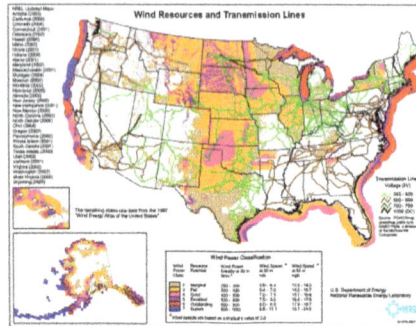

Map of available wind power over the United States. Color codes indicate wind power density class

As a general rule, economic wind generators require windspeed of 16 km/h (10 mph) or greater. An ideal location would have a near constant flow of non-turbulent wind throughout the year, with a minimum likelihood of sudden powerful bursts of wind. An important factor of turbine siting is also access to local demand or transmission capacity.

Usually sites are screened on the basis of a wind atlas, and validated with wind measurements. Meteorological wind data alone is usually not sufficient for accurate siting of a large wind power project. Collection of site specific data for wind speed and direction is crucial to determining site potential in order to finance the project. Local winds are often monitored for a year or more, and detailed wind maps constructed before wind generators are installed.

The wind blows faster at higher altitudes because of the reduced influence of drag. The increase in velocity with altitude is most dramatic near the surface and is affected by topography, surface roughness, and upwind obstacles such as trees or buildings. Typically, the increase of wind speeds with increasing height follows a wind profile power law, which predicts that wind speed rises proportionally to the seventh root of altitude. Doubling the altitude of a turbine, then, increases the expected wind speeds by 10%, and the expected power by 34%.

In general, a distance of 7D (7 × Rotor Diameter of the Wind Turbine) is set between each turbine in a fully developed wind farm, but micrositing optimizes placement, particularly in hilly areas. Individual turbines are interconnected with a medium voltage (usually 34.5 kV) power collection system and communications network. At a substation, this medium-voltage electric current is increased in voltage with a transformer for connection to the high voltage transmission system. Construction of a land-based wind farm requires installation of the collector system and substation, and possibly access roads to each turbine site.

First wind farm consisting of 7.5 megawatt (MW) Enercon E-126 turbines (Estinnes, Belgium, 20 July 2010), two months before completion; note the 2-part blades.

Onshore Installations

Part of the Biglow Canyon Wind Farm, Oregon, United States with a turbine under construction

Wind turbines at the Jepirachí Eolian Park in La Guajira, Colombia.

The world's first wind farm was 0.6 MW, consisting of 20 wind turbines rated at 30 kilowatts each, installed on the shoulder of Crotched Mountain in southern New Hampshire in December 1980.

World's largest onshore wind farms		
Wind farm	Current capacity (MW)	Country
Gansu Wind Farm	6,800	China
Zhang Jiakou	3,000	China
Urat Zhongqi, Bayannur City	2,100	China
Hami Wind Farm	2,000	China
Damao Qi, Baotou City	1,600	China
Alta (Oak Creek-Mojave)	1,320	United States
Jaisalmer Wind Park	1,064	India
Hongshagang, Town, Minqin County	1,000	China
Kailu, Tongliao	1,000	China
Chengde	1,000	China
Buffalo Gap Wind Farm	523.3	United States
Capricorn Ridge Wind Farm	662.5	United States
Dabancheng Wind Farm	500	China
Fântânele-Cogealac Wind Farm	600	Romania
Fowler Ridge Wind Farm	599.8	United States
Horse Hollow Wind Energy Center	735.5	United States
Meadow Lake Wind Farm	500	United States
Panther Creek Wind Farm	458	United States
Roscoe Wind Farm	781.5	United States
Shepherds Flat Wind Farm	845	United States
Sweetwater Wind Farm	585.3	United States
Whitelee Wind Farm	539	Scotland, U.K

Onshore turbine installations in hilly or mountainous regions tend to be on ridgelines generally three kilometres or more inland from the nearest shoreline. This is done to exploit the topographic acceleration as the wind accelerates over a ridge. The additional wind speeds gained in this way can increase energy produced because more wind goes through the turbines. The exact position of each turbine matters, because a difference of 30m could potentially double output. This careful placement is referred to as 'micro-siting'.

Offshore Installations

Offshore wind turbines near Copenhagen, Denmark.

Europe is the leader in offshore wind energy, with the first offshore wind farm (Vindeby) being installed in Denmark in 1991. As of 2010, there are 39 offshore wind farms in waters off Belgium, Denmark, Finland, Germany, Ireland, the Netherlands, Norway, Sweden and the United Kingdom, with a combined operating capacity of 2,396 MW. More than 100 GW (or 100,000 MW) of offshore projects are proposed or under development in Europe. The European Wind Energy Association has set a target of 40 GW installed by 2020 and 150 GW by 2030.

As of September 2013, The London Array in United Kingdom is the largest offshore wind farm in the world at 1,000 MW, followed by Greater Gabbard wind farm (504 MW) also in the UK.

The world's 10 largest offshore wind farms				
Wind farm	Capacity (MW)	Country	Turbines & model	Commissioned
London Array	630	United Kingdom	175 × Siemens SWT-3.6	2013
Greater Gabbard wind farm	504	United Kingdom	140 × Siemens SWT-3.6	2012
Anholt	400	Denmark	111 × Siemens 3.6-120	2013
BARD Offshore 1	400	Germany	80 × BARD 5.0	2013
Walney	367	United Kingdom	102 × Siemens SWT-3.6	2012
Thorntonbank	325	Belgium	6 × 5MW REpower and 48 × 6.15MW REpower	2013
Sheringham Shoal	315	United Kingdom	88 × Siemens 3.6-107	2012
Thanet	300	United Kingdom	100 × Vestas V90-3MW	2010
Lincs	270	United Kingdom	75 × 3.6MW	2013
Horns Rev II	209	Denmark	91 × Siemens 2.3–93	2009

Offshore wind turbines are less obtrusive than turbines on land, as their apparent size and noise is mitigated by distance. Because water has less surface roughness than land (especially deeper water), the average wind speed is usually considerably higher over open water. Capacity factors (utilisation rates) are considerably higher than for onshore locations.

The province of Ontario in Canada is pursuing several proposed locations in the Great Lakes, including the suspended Trillium Power Wind 1 approximately 20 km from shore and over 400 MW in size. Other Canadian projects include one on the Pacific west coast.

As of 2010, there are no offshore wind farms in the United States. However, projects are under development in wind-rich areas of the East Coast, Great Lakes, and Pacific coast; the Block Island Wind Farm is scheduled for 2016.

Installation and service / maintenance of off-shore wind farms are a specific challenge for technology and economic operation of a wind farm. As of 2015, there are 20 jackup vessels for lifting components, but few can lift sizes above 5MW. Service vessels have to be operated nearly 24/7 (availability higher than 80% of time) to get sufficient amortisation from the wind turbines. Therefore, special fast service vehicles for installation (like Wind Turbine Shuttle) as well as for maintenance (including heave compensation and heave compensated working platforms to allow the service staff to enter the wind turbine also at difficult weather conditions) are required. So-called inertial and optical based Ship Stabilization and Motion Control systems (iSSMC) are used for that.

Experimental and Proposed Wind Farms

There exist also some wind parks, which were mainly built for testing wind turbines. In such wind parks, there is usually from each type to be tested only a single wind turbine. Such parks have usually at least one meteorological tower. An example of an experimental wind park is Østerild Wind Turbine Test Field.

For some time, airborne wind farms have been discussed. Airborne wind farm is a group of airborne wind energy systems in the same location, connected to the grid in the same point.

By Region

Australia

The Australian Canunda Wind Farm, South Australia at sunrise

Large operational wind farms in Australia: September 2012			
Wind farm	Installed capacity (MW)	Developer	State
Collgar Wind Farm	206	UBS Investment Bank & Retail Employees Superannuation Trust	Western Australia
Capital Wind Farm	140.7	Infigen Energy	New South Wales
Hallett Group	298	AGL Energy	South Australia
Lake Bonney Wind Farm	278	Infigen Energy	South Australia
Portland Group	132		Victoria
Waubra Wind Farm	192	Acciona Energy & ANZ Infrastructure Services	Victoria
Woolnorth Wind Farm	140	Roaring 40s & Hydro Tasmania	Tasmania

Canada

The Pubnico Wind Farm taken from Beach Point, Lower East Pubnico, Nova Scotia

Large wind farms in Canada			
Name	Capacity (MW)	Location	Province
Anse-à-Valleau Wind Farm	100	Gaspé	Quebec
Caribou Wind Park	99	70 km west of Bathurst	New Brunswick
Bear Mountain Wind Park	120	Dawson Creek	British Columbia
Centennial Wind Power Facility	150	Swift Current	Saskatchewan
Enbridge Ontario Wind Farm	181	Kincardine	Ontario
Erie Shores Wind Farm	99	Port Burwell	Ontario
Jardin d'Eole Wind Farm	127	Saint-Ulric	Quebec
Kent Hills Wind Farm	96	Riverside-Albert	New Brunswick
Melancthon EcoPower Centre	199	Melancthon	Ontario
Port Alma Wind Farm	101	Chatham-Kent	Ontario
Chatham Wind Farm	101	Chatham-Kent	Ontario
Prince Township Wind Farm	189	Sault Ste. Marie	Ontario

Large wind farms in Canada			
Name	Capacity (MW)	Location	Province
St. Joseph Wind Farm	138	Montcalm	Manitoba
St. Leon Wind Farm	99	St. Leon	Manitoba
Wolfe Island Wind Project	197	Frontenac Islands	Ontario

China

Wind farm in Xinjiang, China

In just five years, China leapfrogged the rest of the world in wind energy production, going from 2,599 MW of capacity in 2006 to 62,733 MW at the end of 2011. However, the rapid growth outpaced China's infrastructure and new construction slowed significantly in 2012.

At the end of 2009, wind power in China accounted for 25.1 gigawatts (GW) of electricity generating capacity, and China has identified wind power as a key growth component of the country's economy. With its large land mass and long coastline, China has exceptional wind resources. Researchers from Harvard and Tsinghua University have found that China could meet all of their electricity demands from wind power by 2030.

By the end of 2008, at least 15 Chinese companies were commercially producing wind turbines and several dozen more were producing components. Turbine sizes of 1.5 MW to 3 MW became common. Leading wind power companies in China were Goldwind, Dongfang Electric, and Sinovel along with most major foreign wind turbine manufacturers. China also increased production of small-scale wind turbines to about 80,000 turbines (80 MW) in 2008. Through all these developments, the Chinese wind industry appeared unaffected by the global financial crisis, according to industry observers.

According to the Global Wind Energy Council, the development of wind energy in China, in terms of scale and rhythm, is absolutely unparalleled in the world. The National People's Congress permanent committee passed a law that requires the Chinese energy companies to purchase all the electricity produced by the renewable energy sector.

European Union

A wind farm in a mountainous area in Galicia, Spain

Wind farm in Lower Saxony, Germany

The European Union has a total installed wind capacity of 93,957 MW. Germany has the third largest capacity in the world (after China and the United States) with an installed capacity was 29,060 MW at the end of 2011, and Spain has 21,674 MW. Italy and France each had between 6,000 and 7,000 MW. By January 2014, the UK installed capacity was 10,495 MW. But energy production can be different from capacity – in 2010, Spain had the highest European wind power production with 43 TWh compared to Germany's 35 TWh.

Europe's largest windfarm is the 'London Array', an off-shore wind farm in the Thames Estuary in the United Kingdom, with a current capacity of 630 MW (and thus the world's largest off-shore wind farm). Other large wind farms in Europe include Fântânele-Cogealac Wind Farm near Constanța, Romania with 600 MW capacity, and Whitelee Wind Farm near Glasgow, Scotland which has a total capacity of 539 MW.

An important limiting factor of wind power is variable power generated by wind farms. In most locations the wind blows only part of the time, which means that there has to be back-up capacity of conventional generating capacity to cover periods that the wind is not blowing. To address this issue it has been proposed to create a "supergrid" to connect national grids together across western Europe, ranging from Denmark across the southern North Sea to England and the Celtic Sea to Ireland, and further south to France and Spain especially in Higueruela which was considered for some time the biggest wind farm in the world. The idea is that by the time a low pressure area has moved away from Denmark to the Baltic Sea the next low appears off the coast of Ireland. Therefore, while it is true that the wind is not blowing everywhere all of the time, it will always be blowing somewhere.

India

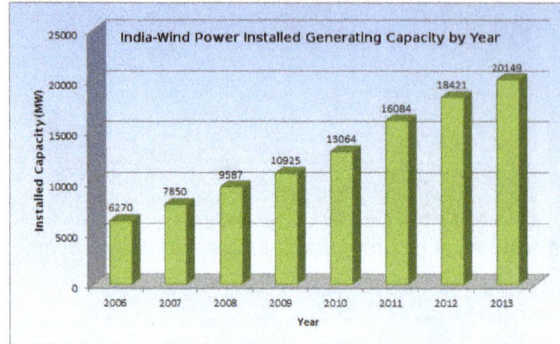

Progress in India's installed wind power generating capacity since 2006

India has the fifth largest installed wind power capacity in the world. As of 31 March 2014, the installed capacity of wind power was 21136.3 MW mainly spread across Tamil Nadu state (7253 MW). Wind power accounts nearly 8.5% of India's total installed power generation capacity, and it generates 1.6% of the country's power.

Japan

Morocco

Morocco has undertaken a vast wind energy program, to support the development of renewable energy and energy efficiency in the country. The Moroccan Integrated Wind Energy Project, spanning over a period of 10 years with a total investment estimated at $3.25 billion, will enable the country to bring the installed capacity, from wind energy, from 280 MW in 2010 to 2000 MW in 2020.

Pakistan

Jhimpir Wind Farm

Pakistan has wind corridors in Jhimpir, Gharo and Keti Bundar in Sindh province and is currently developing wind power plants in Jhimpir and Mirpur Sakro (District Thatta). The government of Pakistan decided to develop wind power energy sources due to problems supplying energy to the southern coastal regions of Sindh and Balochistan. The Zorlu Energy Putin Power Plant is the first wind power plant in Pakistan. The wind farm is being developed in Jhimpir, by Zorlu Energy Pakistan the local subsidiary of a Turkish company. The total cost of project is $136 million. Completed in 2012, it has a total capacity of around 56MW. Fauji Fertilizer Company Energy Limited, has build a 49.5 MW wind Energy Farm at Jhimpir. Contract of supply of mechanical design was awarded to Nordex and Descon Engineering Limited. Nordex a German wind turbine manufacturer. In the end of 2011 49.6 MW will be completed.Pakistani Govt. also has issued LOI of 100 MW Wind power plant to FFCEL. Pakistani Govt. has plans to achieve electric power up to 2500 MW by the end of 2015 from wind energy to bring down energy shortage.

Currently four wind farms are operational (Fauji Fertilizer 49.5 MW (subsidiary of Fauji Foundation), Three Gorges 49.5 MW, Zorlu Energy Pakistan 56 MW, Sapphire Wind Power Co Ltd 52.6 MW) and six are under construction phase (Master Wind Energy Ltd 52.6 MW, Sachal Energy Development Ltd 49.5 MW, Yunus Energy Ltd 49.5 MW, Gul Energy 49.5 MW, Metro Energy 49.5 MW, Tapal Energy) and expected to achieve COD in 2017.

In Gharo wind corridor, two wind farms (Foundation Energy 1 & II each 49.5 MW) are operational while two wind farms Tenaga Generasi Ltd 49.5 MW and HydroChina Dawood Power Pvt Ltd 49.5 are under construction and expected to achieve COD in 2017.

According to a USAID report, Pakistan has the potential of producing 150,000 megawatts of wind energy, of which only the Sindh corridor can produce 40,000 megawatts.

The Philippines

Philippines has the first windfarm in Southeast Asia. Located Northern part of the countries' biggest island Luzon, alongside the seashore of Municipality of Bangui, Province of Ilocos Norte.

The wind farm uses 20 units of 70-metre (230 ft) high Vestas V82 1.65 MW wind turbines, arranged on a single row stretching along a nine-kilometer shoreline off Bangui Bay, facing the West Philippine Sea.

Phase I of the NorthWind power project in Bangui Bay consists of 15 wind turbines, each capable of producing electricity up to a maximum capacity of 1.65 MW, for a total of 24.75 MW. The 15 on-shore turbines are spaced 326 metres (1,070 ft) apart, each 70 metres (230 ft) high, with 41 metres (135 ft) long blades, with a rotor diameter of 82 metres (269 ft) and a wind swept area of 5,281 square metres (56,840 sq ft).

Phase II, was completed on August 2008, and added 5 more wind turbines with the same capacity, and brought the total capacity to 33 MW. All 20 turbines describes a graceful arc reflecting the shoreline of Bangui Bay, facing the West Philippine Sea.

Sri Lanka

Sri Lanka has received funding from the Asian Development Bank amounting to $300 million to invest in renewable energies. From this funding as well as $80 million from the Sri Lankan Government and $60 million from France's Agence Française de Développement, Sri Lanka is building two 100MW wind farms from 2017 due to be completed by late 2020 in Northern Sri Lanka.

South Africa

Turbines at the Gouda Wind Facility just outside the town of Gouda, South Africa.

As of September 2015 a number of sizable wind farms have been constructed in South Africa mostly in the Western Cape region. These include the 100 MW Sere Wind Farm and the 138 MW Gouda Wind Facility.

Most future wind farms in South Africa are earmarked for locations along the Eastern Cape coastline. Eskom has constructed one small scale prototype windfarm at Klipheuwel in the Western Cape and another demonstrator site is near Darling with phase 1 completed. The first commercial wind farm, Coega Wind Farm in Port Elisabeth, was developed by the Belgian company Electrawinds.

Power plant	Province	Date commissioned	Installed Capacity (Megawatt)	Status	Coordinates
Coega Wind Farm	Eastern Cape	2010	1.8 (45)	Operational	33°45′16″S 25°40′30″E33.75444° S 25.67500°E
Darling Wind Farm	Western Cape	2008	5.2 (13.2)	Operational	33°19′55″S 18°14′38″E33.33195°S 18.24378°E
Klipheuwel Wind Farm (af)	Western Cape	2002	3.16	Operational (Prototype/ Research)	33°41′43″S 18°43′30″E33.69539°S 18.72512°E
Sere Wind Farm	Western Cape	2014	100	Operational	31°32′S 18°17′E31.53°S 18.29°E

Gouda Wind Facility	Western Cape	2015	138	Operational	33°17′S 19°03′E33.29°S 19.05°E

United States

Brazos Wind Farm in the plains of West Texas

U.S. wind power installed capacity in 2012 exceeded 51,630 MW and supplies 3% of the nation's electricity.

New installations place the U.S. on a trajectory to generate 20% of the nation's electricity by 2030 from wind energy. Growth in 2008 channeled some $17 billion into the economy, positioning wind power as one of the leading sources of new power generation in the country, along with natural gas. Wind projects completed in 2008 accounted for about 42% of the entire new power-producing capacity added in the U.S. during the year.

At the end of 2008, about 85,000 people were employed in the U.S. wind industry, and GE Energy was the largest domestic wind turbine manufacturer. Wind projects boosted local tax bases and revitalized the economy of rural communities by providing a steady income stream to farmers with wind turbines on their land. Wind power in the U.S. provides enough electricity to power the equivalent of nearly 9 million homes, avoiding the emissions of 57 million tons of carbon each year and reducing expected carbon emissions from the electricity sector by 2.5%.

Texas, with 10,929 MW of capacity, has the most installed wind power capacity of any U.S. state, followed by California with 4,570 MW and Iowa with 4,536 MW. The Alta Wind Energy Center (1,020 MW) in California is the nation's largest wind farm in terms of capacity. Altamont Pass Wind Farm is the largest wind farm in the U.S. in terms of the number of individual turbines.

Criticism

Public perception is that renewable energies such as wind, solar, biomass and geothermal are having a significant positive impact on global warming. All of these sources combined only supplied 1.3% of global energy in 2013 as 8 billion tonnes of coal was burned annually.

One of the biggest factors inhibiting wind farm construction is human opposition. A study has shown "turbine placement close to residents may heighten their uncertainty and concern of the wind turbines and overshadow any positive inclinations towards the development."

Wind farm development is affected by the emphasis being primarily placed on the domain of landscape assessment and environmental impact when seeking farm sites. The viability and efficiency of the wind farm are barely touched upon, instead falling to the developer. For example, Sturge et al. of the University of Sheffield wrote that in many countries where wind energy is becoming popular, engineering aspects, specifically energy yield are not being taken into consideration, either by the public or in the process of planning consent for wind farm development. As energy is the main purpose of wind farms, a lack of attention given to the subject could be detrimental to the general acceptance of wind farms.

Environmental Impact

Livestock ignore wind turbines, and continue to graze as they did before wind turbines were installed.

Compared to the environmental impact of traditional energy sources, the environmental impact of wind power is relatively minor. Wind power consumes no fuel, and emits no air pollution, unlike fossil fuel power sources. The energy consumed to manufacture and transport the materials used to build a wind power plant is equal to the new energy produced by the plant within a few months. While a wind farm may cover a large area of land, many land uses such as agriculture are compatible, with only small areas of turbine foundations and infrastructure made unavailable for use.

There are reports of bird and bat mortality at wind turbines as there are around other artificial structures. The scale of the ecological impact may or may not be significant, depending on specific circumstances. The estimated number of bird deaths caused by wind turbines is between 140,000 and 328,000, whereas deaths caused by domestic cats are estimated to be between 1.3 and 4.0 billion birds each year and over 100 million birds are killed each year by impact with windows. Prevention and mitigation of wildlife fatalities, and protection of peat bogs, affect the siting and operation of wind turbines.

Human Health

There have been multiple scientific, peer-reviewed studies into wind farm noise, which have concluded that infrasound from wind farms is not a hazard to human health and there is no verifiable evidence for 'Wind Turbine Syndrome', although some suggest further research might still be useful.

A 2007 report by the U.S. National Research Council noted that noise produced by wind turbines is generally not a major concern for humans beyond a half-mile or so. Low-frequency vibration and its effects on humans are not well understood and sensitivity to such vibration resulting from wind-turbine noise is highly variable among humans. There are opposing views on this subject, and more research needs to be done on the effects of low-frequency noise on humans.

In a 2009 report about "Rural Wind Farms", a Standing Committee of the Parliament of New South Wales, Australia, recommended a minimum setback of two kilometres between wind turbines and neighbouring houses (which can be waived by the affected neighbour) as a precautionary approach.

A 2014 paper suggests that the 'Wind Turbine Syndrome' is mainly caused by the nocebo effect and other psychological mechanisms.

Effect on Power Grid

Utility-scale wind farms must have access to transmission lines to transport energy. The wind farm developer may be obliged to install extra equipment or control systems in the wind farm to meet the technical standards set by the operator of a transmission line. The company or person that develops the wind farm can then sell the power on the grid through the transmission lines and ultimately chooses whether to hold on to the rights or sell the farm or parts of it to big business like GE, for example.

Ground Radar Interference

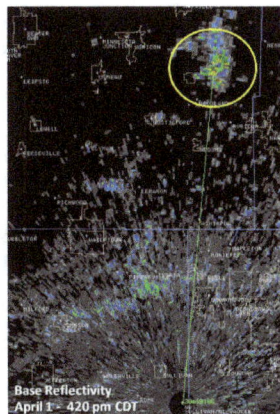

Wind farm interference (in yellow circle) on radar map

Wind farms can interfere with ground radar systems used for military, weather and air traffic control. The large, rapidly moving blades of the turbines can return signals to the radar that can be mistaken as an aircraft or weather pattern. Actual aircraft and weather patterns around wind farms can be accurately detected, as there is no fundamental physical constraint preventing that. But aging radar infrastructure is significantly challenged with the task. The US military is using wind turbines on some bases, including Barstow near the radar test facility.

Effects

The level of interference is a function of the signal processors used within the radar, the speed of the aircraft and the relative orientation of wind turbines/aircraft with respect to the radar. An aircraft flying above the wind farm's turning blades could become impossible to detect because the blade tips can be moving at nearly aircraft velocity. Studies are currently being performed to determine the level of this interference and will be used in future site planning. Issues include masking (shadowing), clutter (noise), and signal alteration. Radar issues have stalled as much as 10,000 MW of projects in USA.

Some very long range radars are not affected by wind farms.

Mitigation

Permanent problem solving include a *non-initiation window* to hide the turbines while still tracking aircraft over the wind farm, and a similar method mitigates the false returns. England's Newcastle Airport is using a short-term mitigation; to "blank" the turbines on the radar map with a software patch. Wind turbine blades using stealth technology are being developed to mitigate radar reflection problems for aviation. As well as stealth windfarms, the future development of infill radar systems could filter out the turbine interference.

In early 2011, the U.S. government awarded a program to build a radar/wind turbine analysis tool. This tool will allow developers to predict the impact of a wind farm on a radar system before construction, thus allowing rearrangement of the turbines or even the entire wind farm to avoid negative impacts on the radar system.

A mobile radar system, the Lockheed Martin TPS-77, has shown in recent tests that it can distinguish between aircraft and wind turbines, and more than 170 TPS-77 radars are in use around the world. In Britain, the Lockheed Martin TPS-77 system was to be installed at Trimingham in Norfolk to remove military objections to a series of offshore wind farms in the North Sea. A second TPS-77 was to be installed in the Scottish Borders, overcoming objections to a 48-turbine wind farm at Fallago.

Radio Reception Interference

There are also reports of negative effects on radio and television reception in wind farm communities. Potential solutions include predictive interference modeling as a component of site selection.

Agriculture

A 2010 study found that in the immediate vicinity of wind farms, the climate is cooler during the day and slightly warmer during the night than the surrounding areas due to the turbulence generated by the blades.

In another study an analysis carried out on corn and soybean crops in the central areas of the United States noted that the microclimate generated by wind turbines improves crops as it prevents the late spring and early autumn frosts, and also reduces the action of pathogenic fungi that grow on the leaves. Even at the height of summer heat, the lowering of 2.5–3 degrees above the crops due to turbulence caused by the blades, can make a difference for the cultivation of corn.

Offshore Wind Power

Wind turbines and electrical substation of *Alpha Ventus supplied by Adwen* in the North Sea

Offshore wind power or offshore wind energy is the use of wind farms constructed offshore, usually on the continental shelf, to harvest wind energy to generate electricity. Stronger wind speeds are available offshore compared to on land, so offshore wind power's contribution in terms of electricity supplied is higher, and NIMBY opposition to construction is usually much weaker. However, offshore wind farms are relatively expensive. At the end of 2014, 3,230 turbines at 84 offshore wind farms across 11 European countries had been installed and grid-connected, making a total capacity of 11,027 MW.

Four offshore wind farms are in the Thames Estuary area: Kentish Flats, Gunfleet Sands, Thanet and London Array. The latter is the largest in the world as of April 2013.

As of 2010 Siemens and Vestas were turbine suppliers for 90% of offshore wind power, while Dong Energy, Vattenfall and E.on were the leading offshore operators. As of 1 January 2016, about 12 gigawatts (GW) of offshore wind power capacity was operational, mainly in Northern Europe, with 3,755 MW of that coming online during 2015. According to BTM Consult, more than 16 GW of additional capacity will be installed before the end of 2014 and the United Kingdom and Germany will become the two leading markets. Offshore wind power capacity is expected to reach a total of 75 GW worldwide by 2020, with significant contributions from China and the United States.

As of 2013 the 630 megawatt (MW) London Array is the largest offshore wind farm in the world, with the 504 (MW) Greater Gabbard wind farm the second largest, followed by the 367 MW Walney Wind Farm. All are off the coast of the UK. These projects will be dwarfed by subsequent wind farms that are in the pipeline, including Dogger Bank at 4,800 MW, Norfolk Bank (7,200 MW), and Irish Sea (4,200 MW). At the end of June 2013 total European combined offshore wind energy capacity was 6,040 MW. UK installed 513.5 MW offshore windpower in the first half year of 2013.

Definition

Offshore wind power refers to the construction of wind farms in bodies of water to generate electricity from wind. Unlike the typical usage of the term "offshore" in the marine industry, offshore wind power includes inshore water areas such as lakes, fjords and sheltered coastal areas, utilizing traditional fixed-bottom wind turbine technologies, as well as deep-water areas utilizing floating wind turbines.

The U.S. National Renewable Energy Laboratory has further defined offshore wind power based on its siting in terms water depth to include shallow water, transitional water, and deep water offshore wind power.

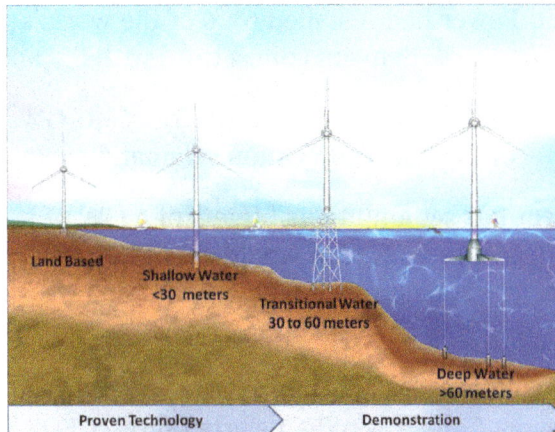

Progression of expected wind turbine evolution to deeper water.

History

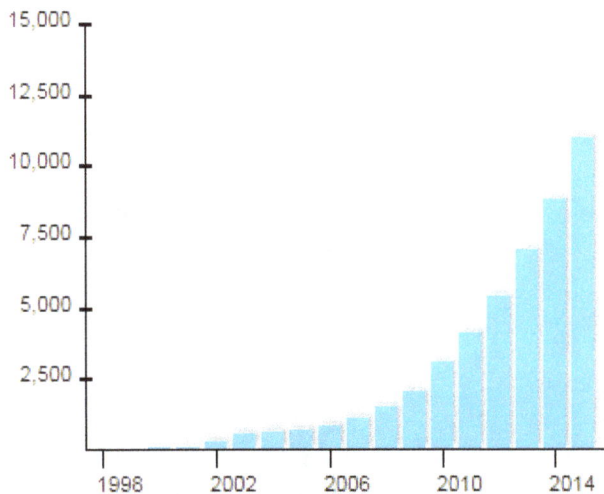

Global cumulative offshore capacity (in megawatts).
Sources: GWEC (2011–2014) and EWEA (1998–2010)

Europe is the world leader in offshore wind power, with the first offshore wind farm (Vindeby) being installed in Denmark in 1991. In 2013, offshore wind power contributed to 1,567 MW of the total 11,159 MW of wind power capacity constructed that year. By January 2014, 69 offshore wind farms had been constructed in Europe with an average annual rated capacity of 482 MW in 2013, and as of January 2014 the United Kingdom has by far the largest capacity of offshore wind farms with 3,681 MW. Denmark is second with 1,271 MW installed and Belgium is third with 571 MW. Germany comes fourth with 520 MW, followed by the Netherlands (247 MW), Sweden (212 MW), Finland (26 MW), Ireland (25 MW), Spain (5 MW), Norway (2 MW) and Portugal (2 MW). By January 2014, the total installed capacity of offshore wind farms in European waters had reached 6,562 MW.

By 2015, Siemens Wind Power had installed 63% of the world's 11 GW offshore wind power capacity; Vestas had 19%, Senvion comes third with 8% and Adwen 6%.

Projections for 2020 calculate a wind farm capacity of 40 GW in European waters, which would provide 4% of the European Union's demand of electricity.

The Chinese government has set ambitious targets of 5 GW of installed offshore wind capacity by 2015 and 30 GW by 2020 that would eclipse capacity in other countries. In May 2014 current capacity of offshore wind power in China was 565 MW.

India is looking at the potential of off-shore wind power plants, with a 100 MW demonstration plant being planned off the coast of Gujarat (2014). In 2013, a group of organizations, led by Global Wind Energy Council (GWEC) started project FOWIND (Facilitating Offshore Wind in India) to identify potential zones for development of off-shore wind power in India and to stimulate R & D activities in this area. In 2014 FOWIND commissioned Center for Study of Science, Technology and Policy (CSTEP) to undertake pre-feasibility studies in eight zones in Tamil Nadu which have been identified as having potential.

Deep Water Offshore Wind Farms

New England Aqua Ventus I, Maine, United States (Planned, 2018-2019)

The patent-pending VolturnUS floating concrete hull technology can support wind turbines in water depths of 45 meters or more, and has the potential to significantly reduce the cost of offshore wind. This technology was developed by the UMaine Advanced Structures and Composites Center.

New England Aqua Ventus I is a two 6 MW turbine (12 MW) floating offshore wind pilot project ~25 km off Maine's coast near Monhegan Island, developed by Maine Aqua Ventus, GP, LLC. This pilot project utilizes the University of Maine's patent-pending VolturnUS, a floating concrete hull technology can support wind turbines in water depths of 45 meters or more. The objective of the pilot is to demonstrate the technology at full scale, allowing floating farms to be built out-of-sight across the US and the world in the 2020s. New England Aqua Ventus I Project partners include Emera Inc., Cianbro Corporation, University of Maine and its Advanced Structures and Composites Center, and DCNS.

This pilot project is planned for deployment in 2018-2019 at the University of Maine's Deepwater Offshore Wind Test Site designated by the State of Maine during the 124th Legislature.

Shallow Water Offshore Wind Farms

World's largest offshore wind farms				
Wind farm	Capacity (MW)	Country	Turbines and model	Commissioned
London Array	630	UK	175 × Siemens SWT-3.6	2012
Greater Gabbard	504	UK	140 × Siemens SWT-3.6	2012
Anholt	400	Denmark	111 × Siemens SWT-3.6-120	2013
BARD Offshore 1	400	Germany	80 × BARD 5.0 turbines	2013
Walney	367	UK	102 × Siemens SWT-3.6	2012
Thorntonbank	325	Belgium	54 × Senvion 6 MW	2013
Sheringham Shoal	317	UK	88 × Siemens 3.6	2013
Thanet	300	UK	100 × Vestas V90-3MW	2010
Meerwind Süd/Ost	288	Germany	80 × Siemens SWT-3.6-120	2014
Lincs	270	UK	75 × Siemens 3.6	2013
Horns Rev II	209	Denmark	91 × Siemens 2.3-93	2009

At the end of 2011, there were 53 European offshore wind farms in waters off Belgium, Denmark, Finland, Germany, Ireland, the Netherlands, Norway, Sweden and the United Kingdom, with an operating capacity of 3,813 MW, while 5,603 MW is under construction. More than 100 GW (or 100,000 MW) of offshore projects are proposed or under development in Europe. The European Wind Energy Association has set a target of 40 GW installed by 2020 and 150 GW by 2030.

As of July 2013, the 175-turbine London Array in the United Kingdom is the largest offshore wind farm in the world with a capacity of 630 MW, followed by Greater Gabbard (504 MW), also in the United Kingdom, Anholt (400 MW) in Denmark, and BARD Offshore 1 (400 MW) in Germany. There are many large offshore wind farms under construction including Gwynt y Môr (576 MW), Borkum West II (400 MW), and West of Duddon Sands (389 MW). Offshore wind farms worth some €8.5 billion ($11.4 billion) were under construction in European waters in 2011. Once completed, they will represent an additional installed capacity of 2,844 MW.

China has two operational offshore wind farms of 131 MW and 101 MW capacity.

Canadian wind power in the province of Ontario is pursuing several proposed locations in the Great Lakes, including the suspended Trillium Power Wind 1 approximately 20 km from shore and over 400 MW in capacity. Other Canadian projects include one on the Pacific west coast.

As of 2015, there are no offshore wind farms in the United States. However, projects are under development in wind-rich areas of the East Coast, Great Lakes, and Pacific coast. In January 2012, a "Smart for the Start" regulatory approach was introduced, designed to expedite the siting process while incorporating strong environmental protections. Specifically, the Department of Interior approved "wind energy areas" off the coast where projects can move through the regulatory approval process more quickly. The first offshore wind farm in the USA is expected to be the 30-megawatt, 5 turbine Block Island Wind Farm which is scheduled to be online in late 2016.

Offshore wind turbines near Copenhagen, Denmark

Economics and Benefits

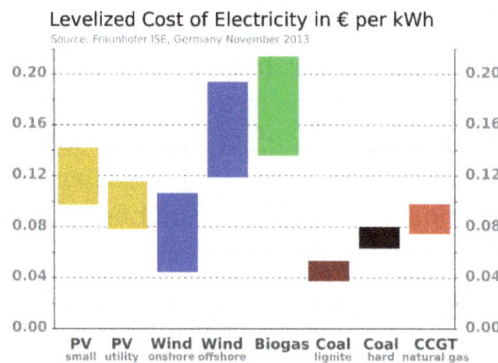

Levelized cost of offshore wind power compared to other sources *(Germany, November 2013)*

Offshore wind power can help to reduce energy imports, reduce air pollution and greenhouse gases (by displacing fossil-fuel power generation), meet renewable electricity standards, and create jobs and local business opportunities. However, according to the US Energy Information Agency, offshore wind power is the most expensive energy generating technology being considered for large scale deployment". The advantage is that the wind is much stronger off the coasts, and unlike wind over the continent, offshore breezes can be strong in the afternoon, matching the time when people are using the most electricity. Offshore turbines can also be "located close to the power-hungry populations along the coasts, eliminating the need for new overland transmission lines".

Most entities and individuals active in offshore wind power believe that prices of electricity will grow significantly from 2009, as global efforts to reduce carbon emissions come into effect. BTM expects cost per kWh to fall from 2014, and that the resource will always be more than adequate in Europe, the United States and China.

The current state of offshore wind power presents economic challenges significantly greater than onshore systems - prices can be in the range of 2.5-3.0 million Euro/MW. The turbine represents just one third to one half of costs in offshore projects today, the rest comes from infrastructure, maintenance, and oversight. Larger turbines with increased energy capture make more economic sense due to the extra infrastructure in offshore systems. Additionally, there are currently no rigorous simulation models of external effects on offshore wind farms, such as boundary layer stability effects and wake effects. This causes difficulties in predicting performance accurately, a critical shortcoming in financing billion-dollar offshore facilities. A report from a coalition of researchers from universities, industry, and government, lays out several things needed in order to bring the costs down and make offshore wind more economically viable:

- Improving wind performance models, including how design conditions and the wind resource are influenced by the presence of other wind farms.

- Reducing the weight of turbine materials

- Eliminating problematic gearboxes

- Turbine load-mitigation controls and strategies

- Turbine and rotor designs to minimize hurricane and typhoon damage

- Economic modeling and optimization of costs of the overall wind farm system, including installation, operations, and maintenance

- Service methodologies, remote monitoring, and diagnostics.

Research and development projects aim to address these issues. One example is the Carbon Trust Offshore Wind Accelerator, a joint industry project, involving nine offshore wind developers, which aims to reduce the cost of offshore wind by 10% by 2015. It has been suggested that innovation at scale could deliver 25% cost reduction in offshore wind by 2020.

In 2011, a Danish energy company claimed that offshore wind turbines are not yet competitive with fossil fuels, but estimates that they will be in 15 years. Until then, state funding and pension funds will be needed. Bloomberg estimates that energy from offshore wind turbines cost 161 euros ($208) per MegaWattHour.

A comprehensive review of the engineering aspects of turbines like the sizes used onshore, including the electrical connections and converters, considers that the industry has in general been overoptimistic about the benefits:costs ratio and concludes that the "offshore wind market doesn't look as if it is going to be big".

In Belfast, the harbour industry is being redeveloped as a hub for offshore windfarm construction, at a cost of about £50m. The work will create 150 jobs in construction, as well as requiring about 1m tonnes of stone from local quarries, which will create hundreds more jobs. "It is the first dedicated harbour upgrade for offshore wind".

The Organisation for Economic Co-operation and Development (OECD) predicts that offshore wind power will grow to 8% of ocean economy by 2030, and that its industry will employ 435,000 people, adding $230 billion of value.

As the first Offshore Windfarms move beyond their initial Warranty periods with the Turbine Equipment Manufacturer an increase in alternative Operations and Maintenance support options is evident. Alternative suppliers of spare parts are entering the market and others are offering niche products and services many of which are focused on improving the power production volumes from these large renewable energy power plants.

As the first offshore wind farms reach their end of life, a demolition industry develops to recycle them at a cost of DKK 2-4 million per MW, to be guaranteed by the owner.

Technical Details

In 2009, the average nameplate capacity of an offshore wind turbine in Europe was about 3 MW, and the capacity of future turbines is expected to increase to 5 MW.

Design Environment

Offshore wind turbines in Fehmarn Belt, the western part of the Baltic Sea between Germany and Denmark (2010)

Offshore wind resource characteristics span a range of spatial and temporal scales and field data on external conditions. For the North Sea, wind turbine energy is around 30 kWh/m^2 of sea area, per year, delivered to grid. The energy per sea area is roughly independent of turbine size. Necessary data includes water depth, currents, seabed, migration, and wave action, all of which drive mechanical and structural loading on potential turbine configurations. Other factors include marine growth, salinity, icing, and the geotechnical characteristics of the sea or lake bed. A number of things are necessary in order to attain the necessary information on these subjects. Existing hardware for these measurements includes Light Detection and Ranging (LIDAR), Sonic

Detection and Ranging (SODAR), radar, autonomous underwater vehicles (AUV), and remote satellite sensing, although these technologies should be assessed and refined, according to a report from a coalition of researchers from universities, industry, and government, supported by the Atkinson Center for a Sustainable Future.

Because of the previous factors, one of the biggest difficulties with offshore wind farms is the ability to predict loads. Analysis must account for the dynamic coupling between translational (surge, sway, and heave) and rotational (roll, pitch, and yaw) platform motions and turbine motions, as well as the dynamic characterization of mooring lines for floating systems. Foundations and substructures make up a large fraction of offshore wind systems, and must take into account every single one of these factors.

Corrosion is also a serious problem and requires detailed design considerations. The prospect of remote monitoring of corrosion looks very promising using expertise utilised by the offshore oil/gas industry and other large industrial plants.

Some of the guidelines for designing offshore wind farms are IEC 61400-3, but in the US several other standards are necessary. In the EU, different national standards are to be straightlined into more cohesive guidelines to lower costs. The standards requires that a loads analysis is based on site-specific external conditions such as wind, wave and currents.

Planning

Offshore turbines require different types of bases for stability, according to the depth of water. To date a number of different solutions exist:

- A monopile (single column) base, six meters in diameter, is used in waters up to 30 meters deep.

- Gravity Base Structures, for use at exposed sites in water 20– 80 m deep.

- Tripod piled structures, in water 20–80 metres deep.

- Tripod suction caisson structures, in water 20-80m deep.

- Conventional steel jacket structures, as used in the oil and gas industry, in water 20-80m deep.

- Floating wind turbines are being developed for deeper water.

The planning and permitting phase can cost more than $10 million, take 5–7 years and have an uncertain outcome. The industry puts pressure on the governments to improve the processes. In Denmark, many of these phases have been deliberately streamlined by authorities in order to minimize hurdles, and this policy has been extended for coastal wind farms with a concept called 'one-stop-shop'. USA introduced a similar model called "Smart from the Start" in 2012.

Maintenance

Turbines are much less accessible when offshore (requiring the use of a service vessel or helicopter for routine access, and a jackup rig for heavy service such as gearbox replacement), and thus reliability is more important than for an onshore turbine. Some wind farms located far from possible onshore bases have service teams living on site in offshore accommodation units.

A maintenance organization performs maintenance and repairs of the components, spending almost all its resources on the turbines. The conventional way of inspecting the blades is for workers to rappel down the blade, taking a day per turbine. Some farms inspect the blades of three turbines per day by photographing them from the monopile through a 600mm lens, avoiding to go up. Others use camera drones.

Because of their remote nature, prognosis and health-monitoring systems on offshore wind turbines will become much more necessary. They would enable better planning just-in-time maintenance, thereby reducing the operations and maintenance costs. According to a report from a coalition of researchers from universities, industry, and government (supported by the Atkinson Center for a Sustainable Future), making field data from these turbines available would be invaluable in validating complex analysis codes used for turbine design. Reducing this barrier would contribute to the education of engineers specializing in wind energy.

Environmental Impact

While the offshore wind industry has grown dramatically over the last several decades, especially in Europe, there is still a great deal of uncertainty associated with how the construction and operation of these wind farms affect marine animals and the marine environment.

Common environmental concerns associated with offshore wind developments include:

- The risk of seabirds being struck by wind turbine blades or being displaced from critical habitats;

- The underwater noise associated with the installation process of driving monopile turbines into the seabed;

- The physical presence of offshore wind farms altering the behavior of marine mammals, fish, and seabirds with attraction or avoidance;

- The potential disruption of the nearfield and farfield marine environment from large offshore wind projects.

The Tethys database provides access to scientific literature and general information on the potential environmental effects of offshore wind energy.

High-altitude Wind Power

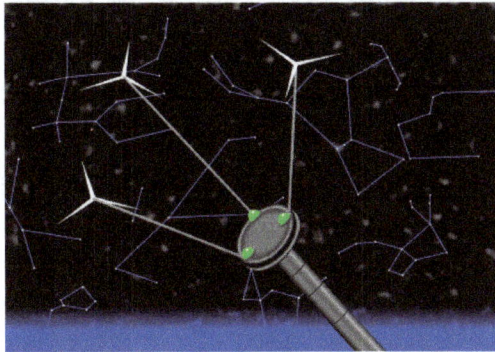

Artist's view of high-altitude wind power

High-altitude wind power (HAWP) is the harnessing of the power of winds high in the sky by use of tether and cable technology. An atlas of the high-altitude wind power resource has been prepared for all points on Earth. A similar atlas of global assessment was developed at Joby Energy. The results were presented at the first annual Airborne Wind Energy Conference held at Stanford University by Airborne Wind Energy Consortium.

Various mechanisms are proposed for capturing the kinetic energy of winds such as kites, kytoons, aerostats, gliders, gliders with turbines for regenerative soaring, sailplanes with turbines, or other airfoils, including multiple-point building- or terrain-enabled holdings. Once the mechanical energy is derived from the wind's kinetic energy, then many options are available for using that mechanical energy: direct traction, conversion to electricity aloft or at ground station, conversion to laser or microwave for power beaming to other aircraft or ground receivers. Energy generated by a high-altitude system may be used aloft or sent to the ground surface by conducting cables, mechanical force through a tether, rotation of endless line loop, movement of changed chemicals, flow of high-pressure gases, flow of low-pressure gases, or laser or microwave power beams.

There are two major scientific articles about jet stream power. Archer & Caldeira calculated that, if the world's entire electrical energy demand were supplied by HAWP, the climatic impact would be negligible. Miller, Gans, & Kleidon claim that the jet streams can generate the total power of only 7.5 TW, and that the climatic impact will be catastrophic.

High-altitude Wind for Power Purposes

Winds at higher altitudes become steadier, more persistent, and of higher velocity. Because power available in wind increases as the cube of velocity (the velocity-cubed law), assuming other parameters remaining the same, doubling a wind's velocity gives $2^3=8$ times the power; tripling the velocity gives $3^3=27$ times the available power. With steadier and more predictable winds, high-altitude wind has an advantage over wind near the ground. Being able to locate HAWP to effective altitudes and using the vertical dimension of airspace for wind farming brings further advantage using high-altitude winds for generating energy.

High-altitude wind generators can be adjusted in height and position to maximize energy return, which is impractical with fixed tower-mounted wind generators.

In each range of altitudes there are altitude-specific concerns being addressed by researchers and developers. As altitude increases, tethers increase in length, the temperature of the air changes, and vulnerability to atmospheric lightning changes. With increasing altitude, exposure to liabilities increase, costs increase, turbulence exposure changes, likelihood of having the system fly in more than one directional strata of winds increases, and the costs of operation changes. HAWP systems that are flown must climb through all intermediate altitudes up to final working altitudes—being at first a low- and then a high- altitude device.

Methods of Capturing Kinetic Energy of High-altitude Winds

Energy can be captured from the wind by kites, kytoons, tethered gliders, tethered sailplanes, aerostats (spherical as well as shaped kytoons), bladed turbines, airfoils, airfoil matrices, drogues, variable drogues, spiral airfoils, Darrieus turbines, Magnus-effect VAWT blimps, multiple-rotor complexes, fabric Jalbert-parafoil kites, uni-blade turbines, flipwings, tethers, bridles, string loops, wafting blades, undulating forms, and piezoelectric materials, and more.

When a scheme's purpose is to propel ships and boats, the objects tether-placed in the wind will tend to have most of the captured energy be in useful tension in the main tether. The aloft working bodies will be operated to maintain useful tension even while the ship is moving. This is the method for powerkiting sports. This sector of HAWP is the most installed method. The folklore is that Benjamin Franklin used the traction method of HAWP. George Pocock was a leader in tugging vehicles by traction.

Controls

HAWP aircraft need to be controlled. Solutions in built systems have control mechanisms variously situated. Some systems are passive, or active, or a mix. When a kite steering unit (KSU) is lofted, the KSU may be robotic and self-contained; a KSU may be operated from the ground via radio-control by a live human operator or by smart computer programs. Some systems have built sensors in the aircraft body that report parameters like position, relative position to other parts. Kite control units (KCU) have involved more than steering; tether reeling speeds and directions can be adjusted in response to tether tensions and needs of the system during a power-generating phase or return-non-power-generating phase. Kite control parts vary widely.

Methods of Converting the Energy

The mechanical energy of the device may be converted to heat, sound, electricity, light, tension, pushes, pulls, laser, microwave, chemical changes, or compression of gases. Traction is a big direct use of the mechanical energy as in tugging cargo ships and kite-boarders. There are several methods of getting the mechanical energy from the wind's kinetic energy. Lighter-than-air (LTA) moored aerostats are employed as lifters of turbines. Heavier-than-air (HTA) tethered airfoils are being used as lifters or turbines themselves. Combinations of LTA and HTA devices in one system are being built and flown to capture HAWP. Even a family of free-flight airborne devices are represented in the literature that capture the kinetic energy of high-altitude winds (beginning with a description in 1967 by Richard Miller in book Without Visible Means of Support) and a contemporary patent application by Dale C. Kramer, soaring sailplane competitor, inventor.

A research on airborne wind turbine technology innovations reveals that the "Kite type AWTs" technique, the most common type, has high scope of growth in the future; it has contributed for about 44% of the total airborne wind energy during 2008–2012. The kite type AWTs extract energy through wind turbines suspended at high altitudes using kites such as multi-tethered kite, kite and dual purpose circular fan,rotary wing kites etc.

Electric Generator Position in a HAWP System

Electricity generation is just one of the optional choices of using captured mechanical energy; however, this option dominates the focus of professionals aiming to supply large amounts of energy to commerce and utilities. A long array of secondary options include tugging water turbines, pumping of water, or compressing air or hydrogen. The position of the electric generator is a distinguishing feature among systems. Flying the generator aloft is done in a variety of ways. Keeping the generator at the mooring region is another large design option. The option in one system of a generator aloft and at the ground station has been used where a small generator operates electronic devices aloft while the ground generator is the big worker to make electricity for significant loads.

Carousel Generator

The "Carousel" configuration several kites fly at a constant height and higher altitudes, pulling in rotation a generator that moves on a wide circular rail. For a large Carousel system, the power obtained can be calculated as of the order of GW, exposing a law that see the power attainable as a function of the diameter raised to the fifth power, while the increment of cost of the generator is linear.

Aerostat-based HAWP

One method of keeping working HAWP systems aloft is to use buoyant aerostats whether or not the electric generator is lifted or left on the ground. The aerostats are usually,
but not always, shaped to achieve a kiting lifting effect. Recharging leaked lifting gas
receives various solutions. In case of productive winds the aerostats are typically blown
down by the aerodynamic drag applied on the wide and unavoidable Reynolds surface
excluding them de facto from the HAWP category.

- W. R. Benoit US Patent 4350897 Lighter than air wind energy conversion system by William R. Benoit, filed Oct 24,1980, and issued: Sep 21, 1982.

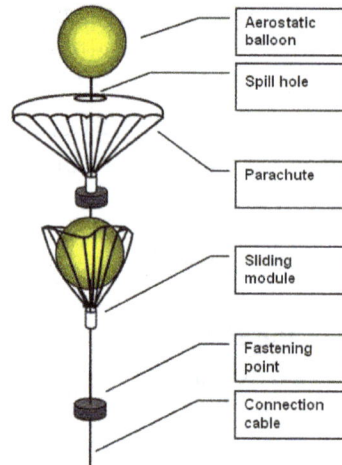

| Aerostatic balloon |
| Spill hole |
| Parachute |
| Sliding module |
| Fastening point |
| Connection cable |

TWIND TECHNOLOGY

- The TWIND system (International patent application PCT/W02010/015720)
  is based on the use of a sail surface elevated by the climbing force of an aerostat-
  ic balloon connected to the ground by a cable used also for energy transmission.
  The wind present at high altitudes creates a horizontal push on the sail which in
  its movement transmits this energy to the ground via the connecting cable. At
  the end of its movement forward, the sail surface is reduced allowing it to move
  upwind with reduced energy waste.

- The Magenn aerostat is a vertical-axis wind turbine held with its axis horizontal by bridling the axis traverse to the wind so that Magnus-effect lift obtains
  during autorotation; the electricity is generated with end-hub generators.

- The LTA Windpower PowerShip uses lift from both an aerostat and wings. It
  operates close to neutral buoyancy and doesn't require a winch. Power is generated by turbines with the propellers on the trailing edge of the wings. The
  system is designed to be able to take off and land unattended.

- Airbine proposes to lift wind turbines into the air by use of aerostats; the electricity would return to ground loads by way of conductive tether.

- Airship power turbine by William J. Mouton, Jr., and David F. Thompson: Their system integrated the turbine within the central portion of a near-toroidal aerostat, like putting a turbine in the hole of an aerostat donut.

- The HAWE system is developed from Tiago Pardal's idea. This system consists of a pumping cycle similar to that of kite systems. In the generation phase, the pulling force increases 5–10 times due to the Magnus effect of a spinning cylinder (aerial platform). Like a kite, the pulling force produced by the aerial platform will unwind the cable and generate electricity on the ground. In the recovery phase it rewinds the cable with no Magnus effect in the aerial platform.

Non-airborne HAWP

Conceptually, two adjacent mountains (natural or terrain-enabled) or artificial buildings or towers (urban or artificial) could have a wind turbine suspended between them by use of cables. When HAWP is cabled between two mountain tops across a valley, the HAWP device is not airborne, but borne up by the cable system. No such systems are known to be in use, though patents teach these methods. When non-cabled bridges are the foundation for holding wind turbines high above the ground, then these are grouped with conventional towered turbines and are outside the intent of HAWP where the tethering an airborne system is foundational.

HAWP Safety

Lightning, aircraft traffic, emergency procedures, system inspections, visibility marking of system parts and its tethers, electrical safety, runaway-wing procedures, over-powering controls, appropriate mooring, and more form the safety environment for HAWP systems.

Challenges of HAWP as an Emerging Industry

There have been several periods of high interest in HAWP before the contemporary activity. The first period had a high focus on pulling carriages over the lands and capturing atmospheric electricity and lightning for human use. The second period was in the 1970s and 1980s when research and investment flourished; a drop in oil price resulted in no significant installations of HAWP. Return on investment (ROI) has been the key parameter; that ROI remains in focus in the current development activity while in the background is the renewable and sustainable energy movement supporting wind power of any kind; but HAWP must compete on ROI with conventional towered solutions.

Early References to HAWP

Early centuries of kiting demonstrated that the kite is a rotary engine that rotates its tether part about its mooring point and causes hands and arms to move because of the energy captured from higher winds into the mechanical device. The tension in the lofted devices performs the work of lifting and pulling body parts and things. Airborne wind energy (AWE) for HAWP was birthed thousands of years ago; naming what happened and developing the implied potentials of tethered aircraft for doing special works is what is occurring in AWE HAWP. What is "low" for some workers is "high" for others.

- 1796 George Pocock used traction mode to travel in vehicles over land roads.

- 1827 George Pocock's book 'The Aeropleustic Art' or 'Navigation in the Air by the Use of Kites or Buoyant Sails' was published. Pocock described use of kites for land and sea travel. The book was republished several times.

- 1833 John Adolphus Etzler saw HAWP blossoming at least for traction.

- 1864? Book's chapter *Kite-Ship* well describes key dynamics of HAWP used for tugging ships by kites. John Gay's: or Work for Boys. Chapter XVIII in the Summer volume.

- 1935 Aloys van Gries stands as a strong early patentee of high-altitude wind power; he taught various kite systems for use in generating electricity in his DE 656194 C patent: Durch Drachen getragene Windkraftmaschine zur Nutzbarmachung von Hoehenwinden

- 1943 Stanley Biszak instructed using potential energy in free-flight for converting ambient winds impacting turbine to drive electric generator to charge batteries.

- 1967 Richard Miller, former editor of Soaring magazine, published book Without Visible Means of Support that describes the feasibility of free-flight coupled non-ground-moored kites to capture differences in wind strata to travel across continents; such HAWP is the subject of Dale C. Kramer's contemporary patent application.

- 1973? Hermann Oberth In the appendix of his book Primer for Those Who Would Govern there are sketches and a photograph of a model of the *Kite Power Station* from the Oberth Museum.

- 1977 April 3, 1977, invention declared. On September 21, 1979, Douglas Selsam notarized his kite-lifted endless chain of airfoils HAWP system, generic type that would later show in Dutch astronaut Wubbo Ockels' device called Ladder-Mill described in a patent of 1997. Douglas Selsam conceived his *Auto-oriented Wind Harnessing Buoyant Aerial Tramway* on April 3, 1977. On the Selsam notarized disclosure of invention was placed a date of Sept. 20, while the notary placed the final signing on Sept. 21, 1979.

- 1979 Professor Bryan Roberts begins giromill gyrocopter-type HAWP wind generator development.

- 1980 Myles Loyd publishes an article on the crosswind kite power.

- 1986 Bryan Roberts' AWE HAWP rotor generates electricity and lifts itself in tethered flight.

- 1992 *Free Rotor* WO/1992020917 Free Rotor by JACK, Colin, Humphry, Bruce (one man). Colin Jack. Colin Bruce. Multi-rotors are treated. Faired tethers are recognized. 1992.

Windmill

The smock mill Goliath in front of the wind farm Growind in Eemshaven in the Netherlands

A windmill is a mill that converts the energy of wind into rotational energy by means of vanes called sails or blades. Centuries ago, windmills usually were used to mill grain, pump water, or both. Thus they often were gristmills, windpumps, or both. The majority of modern windmills take the form of wind turbines used to generate electricity, or windpumps used to pump water, either for land drainage or to extract groundwater.

Windmills in Antiquity

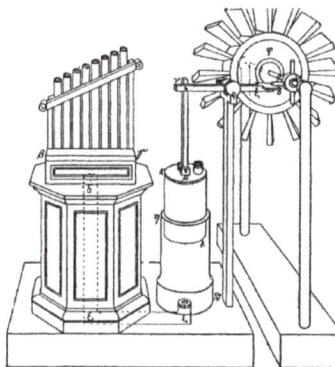

Heron's wind-powered organ

The windwheel of the Greek engineer Heron of Alexandria in the first century is the earliest known instance of using a wind-driven wheel to power a machine. Another early example of a wind-driven wheel was the prayer wheel, which has been used in Tibet and China since the fourth century. It has been claimed that the Babylonian emperor Hammurabi planned to use wind power for his ambitious irrigation project in the seventeenth century BCE.

Horizontal Windmills

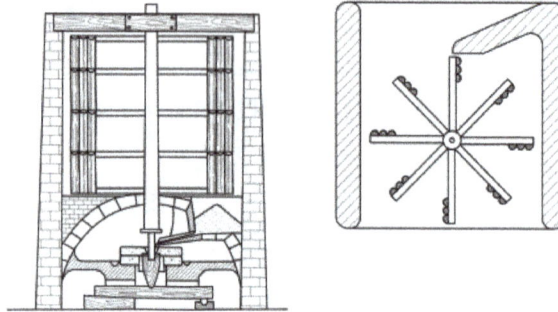

The Persian horizontal windmill

Hooper's Mill, Margate, Kent, an eighteenth-century European horizontal windmill

The first practical windmills had sails that rotated in a horizontal plane, around a vertical axis. According to Ahmad Y. al-Hassan, these panemone windmills were invented in eastern Persia as recorded by the Persian geographer Estakhri in the ninth century. The authenticity of an earlier anecdote of a windmill involving the second caliph Umar (AD 634–644) is questioned on the grounds that it appears in a tenth-century document. Made of six to 12 sails covered in reed matting or cloth material, these windmills were used to grind grain or draw up water, and were quite different from the later European vertical windmills. Windmills were in widespread use across the Middle East and Central Asia, and later spread to China and India from there.

A similar type of horizontal windmill with rectangular blades, used for irrigation, can also be found in thirteenth-century China (during the Jurchen Jin Dynasty in the north), introduced by the travels of Yelü Chucai to Turkestan in 1219.

Horizontal windmills were built, in small numbers, in Europe during the 18th and nineteenth centuries, for example Fowler's Mill at Battersea in London, and Hooper's Mill at Margate in Kent. These early modern examples seem not to have been directly influenced by the horizontal windmills of the Middle and Far East, but to have been independent inventions by engineers influenced by the Industrial Revolution.

Vertical Windmills

Due to a lack of evidence, debate occurs among historians as to whether or not Middle Eastern horizontal windmills triggered the original development of European windmills. In northwestern Europe, the horizontal-axis or vertical windmill (so called due to the plane of the movement of its sails) is believed to date from the last quarter of the twelfth century in the triangle of northern France, eastern England and Flanders.

The earliest certain reference to a windmill in Europe (assumed to have been of the vertical type) dates from 1185, in the former village of Weedley in Yorkshire which was located at the southern tip of the Wold overlooking the Humber estuary. A number of earlier, but less certainly dated, twelfth-century European sources referring to windmills have also been found. These earliest mills were used to grind cereals.

Post Mill

The evidence at present is that the earliest type of European windmill was the post mill, so named because of the large upright post on which the mill's main structure (the "body" or "buck") is balanced. By mounting the body this way, the mill is able to rotate to face the wind direction; an essential requirement for windmills to operate economically in north-western Europe, where wind directions are variable. The body contains all the milling machinery. The first post mills were of the sunken type, where the post was buried in an earth mound to support it. Later, a wooden support was developed called the trestle. This was often covered over or surrounded by a roundhouse to protect the trestle from the weather and to provide storage space. This type of windmill was the most common in Europe until the nineteenth century, when more powerful tower and smock mills replaced them.

Hollow-post Mill

In a hollow-post mill, the post on which the body is mounted is hollowed out, to accommodate the drive shaft. This makes it possible to drive machinery below or outside the body while still being able to rotate the body into the wind. Hollow-post mills driving scoop wheels were used in the Netherlands to drain wetlands from the fourteenth century onwards.

Tower Mill

Tower mills in Spain

By the end of the thirteenth century, the masonry tower mill, on which only the cap is rotated rather than the whole body of the mill, had been introduced. The spread of tower mills came with a growing economy that called for larger and more stable sources of power, though they were more expensive to build. In contrast to the post mill, only the cap of the tower mill needs to be turned into the wind, so the main structure can be made much taller, allowing the sails to be made longer, which enables them to provide useful work even in low winds. The cap can be turned into the wind either by winches or gearing inside the cap or from a winch on the tail pole outside the mill. A method of keeping the cap and sails into the wind automatically is by using a fantail, a small windmill mounted at right angles to the sails, at the rear of the windmill. These are also fitted to tail poles of post mills and are common in Great Britain and English-speaking countries of the former British Empire, Denmark, and Germany but rare in other places. Around some parts of the Mediterranean Sea, tower mills with fixed caps were built because the wind's direction varied little most of the time.

Smock Mill

Two smock mills with a stage in Greetsiel, Germany

The smock mill is a later development of the tower mill, where the tower is replaced by a wooden framework, called the "smock." The smock is commonly of octagonal plan, though examples with more, or fewer, sides exist. The smock is thatched, boarded or covered by other materials, such as slate, sheet metal, or tar paper. The lighter construction in comparison to tower mills make smock mills practical as drainage mills as these often had to be built in areas with unstable subsoil. Having originated as a drainage mill, smock mills are also used for a variety of purposes. When used in a built-up area it is often placed on a masonry base to raise it above the surrounding buildings.

Mechanics

Sails

Windmill in Kuremaa, Estonia

Common sails consist of a lattice framework on which a sailcloth is spread. The miller can adjust the amount of cloth spread according to the amount of wind available and power needed. In medieval mills, the sailcloth was wound in and out of a ladder type arrangement of sails. Postmedieval mill sails had a lattice framework over which the sailcloth was spread, while in colder climates, the cloth was replaced by wooden slats, which were easier to handle in freezing conditions. The jib sail is commonly found in Mediterranean countries, and consists of a simple triangle of cloth wound round a spar.

In all cases, the mill needs to be stopped to adjust the sails. Inventions in Great Britain in the late eighteenth and nineteenth centuries led to sails that automatically adjust to the wind speed without the need for the miller to intervene, culminating in patent sails invented by William Cubitt in 1807. In these sails, the cloth is replaced by a mechanism of connected shutters.

In France, Pierre-Théophile Berton invented a system consisting of longitudinal wooden slats connected by a mechanism that lets the miller open them while the mill is turning. In the twentieth century, increased knowledge of aerodynamics from the development of the airplane led to further improvements in efficiency by German engineer Bilau and several Dutch millwrights.

The majority of windmills have four sails. Multiple-sailed mills, with five, six or eight sails, were built in Great Britain (especially in and around the counties of Lincolnshire and Yorkshire), Germany, and less commonly elsewhere. Earlier multiple-sailed mills are found in Spain, Portugal, Greece, parts of Romania, Bulgaria, and Russia. A mill with an even number of sails has the advantage of being able to run with a damaged sail and the one opposite removed without resulting in an unbalanced mill.

De Valk windmill in mourning position following the death of Queen Wilhelmina of the Netherlands in 1962

In the Netherlands the stationary position of the sails, i.e. when the mill is not working, has long been used to give signals. A slight tilt of the sails before the main building signals joy, while a tilt after the building signals mourning. Across the Netherlands, windmills were placed in mourning position in honor of the Dutch victims of the 2014 Malaysian Airlines Flight 17 shootdown.

Machinery

Gears inside a windmill convey power from the rotary motion of the sails to a mechanical device. The sails are carried on the horizontal windshaft. Windshafts can be wholly made of wood, or wood with a cast iron poll end (where the sails are mounted) or entirely of cast iron. The brake wheel is fitted onto the windshaft between the front and rear bearing. It has the brake around the outside of the rim and teeth in the side of the rim which drive the horizontal gearwheel called wallower on the top end of the vertical upright shaft. In grist mills, the great spur wheel, lower down the upright shaft, drives one or more stone nuts on the shafts driving each millstone. Post mills sometimes have a head and/or tail wheel driving the stone nuts directly, instead of the spur gear arrangement. Additional gear wheels drive a sack hoist or other machinery. The machinery differs if the windmill is used for other applications than milling grain. A drainage mill uses another set of gear wheels on the bottom end of the upright shaft to drive a scoop wheel or Archimedes' screw. Sawmills use a crankshaft to provide a reciprocating motion to the saws. Windmills have been used to power many other industrial processes, including papermills, threshing mills, and to process oil seeds, wool, paints and stone products.

Littlefield, Texas, claims the world's tallest windmill.

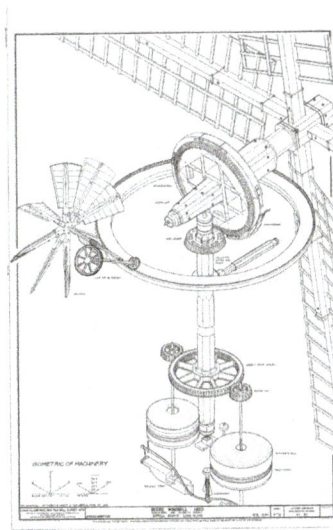

An isometric drawing of the machinery of the Beebe Windmill

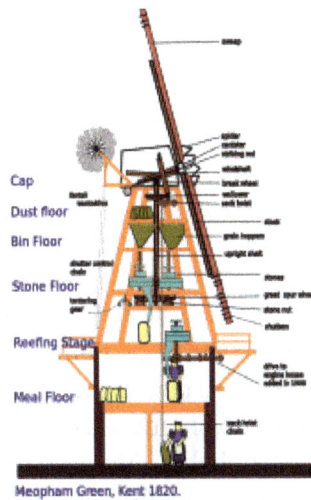

Diagram of the smock mill at Meopham, Kent

The total number of wind-powered mills in Europe is estimated to have been around 200,000 at its peak, which is modest compared to some 500,000 waterwheels. Wind-mills were applied in regions where there was too little water, where rivers freeze in winter and in flat lands where the flow of the river was too slow to provide the required power. With the coming of the industrial revolution, the importance of wind and water as primary industrial energy sources declined and were eventually replaced by steam (in steam mills) and internal combustion engines, although windmills continued to be built in large numbers until late in the nineteenth century. More recently, windmills have been preserved for their historic value, in some cases as static exhibits when the antique machinery is too fragile to put in motion, and in other cases as fully working mills.

Don Quijote being struck by a windmill, illustration by Paul Gustave Louis Christophe Doré.

Of the 10,000 windmills in use in the Netherlands around 1850, about 1000 are still standing. Most of these are being run by volunteers, though some grist mills are still operating commercially. Many of the drainage mills have been appointed as backup to the modern pumping stations. The Zaan district has been said to have been the first industrialized region of the world with around 600 operating wind-powered industries by the end of the eighteenth century. Economic fluctuations and the industrial revolution had a much greater impact on these industries than on grain and drainage mills, so only very few are left.

Construction of mills spread to the Cape Colony in the seventeenth century. The early tower mills did not survive the gales of the Cape Peninsula, so in 1717, the Heeren XVII sent carpenters, masons, and materials to construct a durable mill. The mill, completed in 1718, became known as the *Oude Molen* and was located between Pinelands Station and the Black River. Long since demolished, its name lives on as that of a Technical school in Pinelands. By 1863, Cape Town could boast 11 mills stretching from Paarden Eiland to Mowbray.

Wind Turbines

Rønland Windpark in Denmark

A group of wind turbines in Zhangjiakou, China

A wind turbine is a windmill-like structure specifically developed to generate electricity. They can be seen as the next step in the development of the windmill. The first wind turbines were built by the end of the nineteenth century by Prof James Blyth in Scotland (1887), Charles F. Brush in Cleveland, Ohio (1887–1888) and Poul la Cour in Denmark (1890s). La Cour's mill from 1896 later became the local powerplant of the village Askov. By 1908 there were 72 wind-driven electric generators in Denmark, ranging from 5 to 25 kW. By the 1930s, windmills were widely used to generate electricity on farms in the United States where distribution systems had not yet been installed, built by companies such as Jacobs Wind, Wincharger, Miller Airlite, Universal Aeroelectric, Paris-Dunn, Airline, and Winpower. The Dunlite Corporation produced turbines for similar locations in Australia.

Forerunners of modern horizontal-axis utility-scale wind generators were the WIME-3D in service in Balaklava USSR from 1931 until 1942, a 100-kW generator on a 30-m (100-ft) tower, the Smith-Putnam wind turbine built in 1941 on the mountain known as Grandpa's Knob in Castleton, Vermont, United States of 1.25 MW and the NASA wind turbines developed from 1974 through the mid-1980s. The development of these 13 experimental wind turbines pioneered many of the wind turbine design technologies in use today, including: steel tube towers, variable-speed generators, composite blade materials, and partial-span pitch control, as well as aerodynamic, structural, and acoustic engineering design capabilities. The modern wind power industry began in 1979 with the serial production of wind turbines by Danish manufacturers Kuriant, Vestas, Nordtank, and Bonus. These early turbines were small by today's standards, with capacities of 20–30 kW each. Since then, commercial turbines have increased greatly in size, with the Enercon E-126 capable of delivering up to 7 MW, while wind turbine production has expanded to many countries.

As the 21st century began, rising concerns over energy security, global warming, and eventual fossil fuel depletion led to an expansion of interest in all available forms of renewable energy. Worldwide, many thousands of wind turbines are now operating, with a total nameplate capacity of 194,400 MW. Europe accounted for 48% of the total in 2009.

Windpumps

Aermotor-style windpump in South Dakota, USA

Windpumps were used to pump water since at least the 9th century in what is now Afghanistan, Iran and Pakistan. The use of wind pumps became widespread across the Muslim world and later spread to China and India. Windmills were later used extensively in Europe, particularly in the Netherlands and the East Anglia area of Great Britain, from the late Middle Ages onwards, to drain land for agricultural or building purposes.

The American windmill, or wind engine, was invented by Daniel Halladay in 1854 and was used mostly for lifting water from wells. Larger versions were also used for tasks such as sawing wood, chopping hay, and shelling and grinding grain. In early California and some other states, the windmill was part of a self-contained domestic water system which included a hand-dug well and a wooden water tower supporting a redwood tank enclosed by wooden siding known as a tankhouse. During the late 19th century steel blades and steel towers replaced wooden construction. At their peak in 1930, an estimated 600,000 units were in use. Firms such as U.S. Wind Engine and Pump Company, Challenge Wind Mill and Feed Mill Company, Appleton Manufacturing Company, Star, Eclipse, Fairbanks-Morse, and Aermotor became the main suppliers in North and South America. These windpumps are used extensively on farms and ranches in the United States, Canada, Southern Africa, and Australia. They feature a large number of blades, so they turn slowly with considerable torque in low winds and are self-regulating in high winds. A tower-top gearbox and crankshaft convert the rotary motion into reciprocating strokes carried downward through a rod to the pump cylinder below. Such mills pumped water and powered feed mills, saw mills, and agricultural machinery.

In Australia, the Griffiths Brothers at Toowoomba manufactured windmills from 1876, with the trade name Southern Cross Windmills in use from 1903. These became an icon of the Australian rural sector by utilizing the water of the Great Artesian Basin.

Windpump

A multi-bladed windpump on a farm in Iowa

A windpump is a type of windmill which is used for pumping water.

De Olifant at Burdaard, Friesland

Windpumps were used to pump water since at least the 9th century in what is now Afghanistan, Iran and Pakistan. The use of wind pumps became widespread across the Muslim world and later spread to China and India. Windmills were later used extensively in Europe, particularly in the Netherlands and the East Anglia area of Great Britain, from the late Middle Ages onwards, to drain land for agricultural or building purposes.

Simon Stevin's work in the *waterstaet* involved improvements to the sluices and spillways to control flooding. Windmills were already in use to pump the water out, but in *Van de Molens* (*On mills*), he suggested improvements, including the idea that the wheels should move slowly, and a better system for meshing of the gear teeth. These improvements increased the efficiency of the windmills used to pump water out of the polders by three times. He received a patent on his innovation in 1586.

Eight- to ten-bladed windmills were used in the Region of Murcia, Spain, to raise water for irrigation purposes. The drive from the windmill's rotor was led down through the tower and back out through the wall to turn a large wheel known as a *noria*. The *noria* supported a bucket chain which dangled down into the well. The buckets were traditionally made of wood or clay. These windmills remained in use until the 1950s, and many of the towers are still standing.

Early immigrants to the New World brought with them the technology of windmills from Europe. On US farms, particularly on the Great Plains, wind pumps were used to pump water from farm wells for cattle. In California and some other states, the windmill was part of a self-contained domestic water system, including a hand-dug well and a redwood water tower supporting a redwood tank and enclosed by redwood siding (tankhouse). The self-regulating farm wind pump was invented by Daniel Halladay in 1854. Eventually, steel blades and steel towers replaced wooden construction, and at their peak in 1930, an estimated 600,000 units were in use, with capacity equivalent to 150 megawatts. Very large lighter wind pumps in Australia directly crank the pump with the rotor of the windmill. Extra back gearing between small rotors for high wind areas and the pump crank prevents trying to push the pump rods down on the downstroke faster than they can fall by gravity. Otherwise pumping too fast leads to the pump rods buckling, making the seal of the stuffing box leak and wearing through the wall of the rising main (UK) or the drop-pipe (US) so all output is lost.

The multi-bladed wind pump or wind turbine atop a lattice tower made of wood or steel hence became, for many years, a fixture of the landscape throughout rural America. These mills, made by a variety of manufacturers, featured a large number of blades so that they would turn slowly with considerable torque in moderate winds and be self-regulating in high winds. A tower-top gearbox and crankshaft converted the rotary motion into reciprocating strokes carried downward through a rod to the pump cylinder below. Today, rising energy costs and improved pumping technology are increasing interest in the use of this once declining technology.

Worldwide Use

The Netherlands is well known for its windmills. Most of these iconic structures situated along the edge of polders are actually windpumps, designed to drain the land. These are particularly important as much of the country lies below sea level.

A working wooden windpump on The Fens in Cambridgeshire, UK

In the UK, the term *windpump* is rarely used, and they are better known as *drainage windmills*. Many of these were built in The Broads and The Fens of East Anglia for the draining of land, but most of them have since been replaced by diesel or electric powered pumps. Many of the original windmills still stand in a derelict state although some have been restored.

Windpumps are used extensively in Southern Africa, Australia, and on farms and ranches in the central plains and Southwest of the United States. In South Africa and Namibia thousands of windpumps are still operating. These are mostly used to provide water for human use as well as drinking water for large sheep stocks.

Kenya has also benefited from the African development of windpump technologies. At the end of the 1970s, the UK NGO Intermediate Technology Development Group provided engineering support to the Kenyan company Bobs Harries Engineering Ltd for the development of the Kijito windpumps. Bobs Harries Engineering Ltd is still manufacturing the Kijito windpumps, and more than 300 of them are operating in the whole of East Africa.

In many parts of the world, a rope pump is being used in conjunction with wind turbines. This easy-to- construct pump works by pulling a knotted rope through a pipe (usually a simple PVC pipe) causing the water to be pulled up into the pipe. This type of pump has become common in Nicaragua and other places.

Construction

To construct a windpump, the bladed rotor needs to be matched to the pump. With non-electric windpumps, high solidity rotors are best used in conjunction with positive displacement (piston) pumps, because single-acting piston pumps need about three times as much torque to start them as to keep them going. Low solidity rotors, on the other hand, are best used with centrifugal pumps, waterladder pumps and chain and washer pumps, where the torque needed by the pump for starting is less than that needed for running at design speed. Low solidity rotors are best used if they are intended to drive an electricity generator; which in turn can drive the pump.

Multi-bladed Windpumps

Wind powered water pump on Oak Park Farm, Shedd, Oregon.

Multi-bladed wind pumps can be found worldwide and are manufactured in the United States, Argentina, China, New Zealand, and South Africa. A 16 ft (4.8 m) diameter wind pump can lift up to 1600 US gallons (about 6.4 metric tons) of water per hour to an elevation of 100 ft with a 15 to 20 mph wind (24–32 km/h). However they take a strong wind to start so they turn over the crank of the piston pump. Wind pumps require little maintenance—usually only a change of gear box oil annually. An estimated 60,000 wind pumps are still in use in the United States. They are particularly attractive for use at remote sites where electric power is not available and maintenance is difficult to provide.

A common multi-bladed windpump usefully pumps with about 4%–8% of the annual windpower passing through the area it sweeps This lower conversion is due to poor load matching between wind rotors and fixed-stroke piston pumps.

Fundamental Problems of Multi-bladed Windpumps

Inefficient Rotor

Derelict water tank with windmill in the background

The main design feature of a multi-bladed rotor is "high starting torque", which is necessary for cranking a piston pump operation. Once started a multi-bladed rotor runs at too high a tipspeed ratio at less than its best efficiency of 30% . On the other hand, modern wind rotors can operate at an aerodynamic efficiency of more than 40% at higher tipspeed ratio for a smaller swirl added and wasted to the wind. But they would need a highly variable stroke mechanism rather than just a crank to piston pump.

Poor Load Matching

A multi-bladed windmill is a mechanical device with a piston pump. Because a piston pump has a fixed stroke, the energy demand of this type of pump is proportional to pump speed only. On the other hand, the energy supply of a wind rotor is proportional to the cube of wind speed. Because of that, a wind rotor runs at over speed (more speed than needed), yielding a loss of aerodynamic efficiency.

A variable stroke would match the rotor speed according to wind speed, functioning like a "variable-speed generator". The flow rate of variable stroke windpump can be increased two times, compared to fixed stroke windpumps at the same wind speed.

More detailed information can be found in the introduction, chapter 2 and the appendix of Optimal Wind Turbine Theory a free e-book.

Cyclic Torque Variation

A piston pump has a very light suction phase, but the upstroke is heavy and puts a big backtorque on a starting rotor when the crank is horizontal and ascending. A counterweight on the crank up in the tower and yawing with the wind direction can at least spread the torque to the crank descent.

Development of Improved Windpumps

Although multi-bladed windpumps are based on proven technology and are widely used, they have the fundamental problems mentioned above and need a practical variable stroke mechanism .

USDA Experiments at Texas

Between 1988 and 1990, a variable stroke windpump was tested at the USDA-Agriculture Research Center-Texas, based on two patented designs (Don E. Avery Patent #4.392.785, 1983 and Elmo G. Harris Patent #617.877, 1899). Control systems of the variable stroke wind pumps were mechanical and hydraulic; however, those experiments did not attract the attention of any windpump manufacturer. After experiments with this variable stroke windpump, research focused on wind-electric water pumping systems; no commercial variable stroke windpump exists yet.

Fluttering Windpumps

Fluttering windpumps have been developed in Canada with a pump stroke varying strongly with amplitude to absorb all the variable power in the wind and to stop the uniblade from swinging too far beyond horizontal from its vertical mean position. They are much lighter and use less material than multiblade windpumps and can pump effectively in lighter wind regimes.]

Turkish Experiments

A Turkish engineer re-designed the variable stroke windpump technology by using modern electronic control equipment. Research began in 2004, with governmental R&D support. The first commercial new generation variable stroke wind pumps have been designed after ten years of R&D. The 30 kW variable stroke windpump design includes a Darrieus-type modern wind rotor, counterbalance and regenerative brake technology.

Combinations

Tjasker

The *tjasker*

In the Netherlands, the *tjasker* is a drainage mill with common sails connected to an Archimedean screw. This is used for pumping water in areas where only a small lift is required. The windshaft sits on a tripod which allows it to pivot. The Archimedean screw lifts water into a collecting ring, where it is drawn off into a ditch at a higher level, thus draining the land.

Thai Windpumps

In Thailand, windpumps are traditionally built on Chinese windpump designs. These pumps are constructed from wire-braced bamboo poles carrying fabric or bamboo-mat sails; a paddle pump or waterladder pump is fixed to a Thai bladed rotor. They are mainly used in salt pans where the water lift required is typically less than 1 metre.

References

- The Ocean Economy in 2030, pp.205-212. OECD iLibrary, 27 April 2016. ISBN 9264251723 . Web read

- Lucas, Adam (2006). Wind, Water, Work: Ancient and Medieval Milling Technology. Brill Publishers. p. 105. ISBN 90-04-14649-0.

- Sathyajith, Mathew (2006). Wind Energy: Fundamentals, Resource Analysis and Economics. Springer Berlin Heidelberg. pp. 1–9. ISBN 978-3-540-30905-5.

- Ahmad Y Hassan, Donald Routledge Hill (1986). Islamic Technology: An illustrated history, p. 54. Cambridge University Press. ISBN 0-521-42239-6.

- Lucas, Adam (2006), Wind, Water, Work: Ancient and Medieval Milling Technology, Brill Publishers, pp. 106–7, ISBN 90-04-14649-0

- Laurence Turner, Roy Gregory (2009). Windmills of Yorkshire. Catrine, East Ayrshire: Stenlake Publishing. p. 2. ISBN 9781840334753.

- History of Wind Energy in Cutler J. Cleveland,(ed) Encyclopedia of Energy Vol.6, Elsevier, ISBN 978-1-60119-433-6, 2007, pp. 421-422

- Erich Hau, Wind turbines: fundamentals, technologies, application, economics, Birkhäuser, 2006 ISBN 3-540-24240-6, page 32, with a photo

- Lucas, Adam (2006), Wind, Water, Work: Ancient and Medieval Milling Technology, Brill Publishers, p. 65, ISBN 90-04-14649-0

- Paul Gipe, Wind Energy Comes of Age, John Wiley and Sons, 1995 ISBN 0-471-10924-X, pages 123-127

- Lucas, Adam (2006), Wind, Water, Work: Ancient and Medieval Milling Technology, Brill Publishers, p. 65, ISBN 90-04-14649-0

- Glenn Cramer; Sheldon Town Councilman (30 October 2009). "Town Councilor regrets High Sheldon Wind Farm (Sheldon, NY)". Retrieved 4 September 2015.

- Broadcast Wind, LLC. "Solutions for the Broadcasting and Wind Energy Industries". Retrieved 4 September 2015.

- "IMPACT OF WIND FARMS ON RADIOCOMMUNICATION SERVICES". TSR (grupo Tratamiento de Señal y Radiocomunicaciones de la UPV/EHU). Retrieved 4 September 2015.

- "Wind in our Sails, A report by the European Wind Energy Association - 2011" (PDF). European Wind Energy Association. 2011. p. 11. Retrieved 27 February 2015.

- R. Srikanth; Sangeetha Kandavel (January 29, 2015). "Tapping the offshore wind". The Hindu. Retrieved 30 April 2015.

- "America Is Finally Getting Its First Offshore Wind Farm. Conservatives Are Trying to Make Sure It's the Last.". New Republic. 14 May 2015. Retrieved 15 May 2015.

- Stiesdal, Henrik. "Pi and turbines - a useful context" Original, in Danish Ingeniøren, 13 March 2015. Accessed: 13 March 2015.

- Clements, Elizabeth. "Historic Turns in The Windmill City". Ferimi News. Office of Science/US Dept of Energy. Retrieved 25 January 2015.

3

Wind Turbine: A Comprehensive Study

Wind turbine is a device that converts wind into electrical power. They are manufactured in a wide range; the range differs from vertical to horizontal axis types. Some of the aspects discussed within this chapter are wind turbine design, airborne wind turbine, vertical axis wind turbine, floating wind turbine and unconventional wind turbines. This chapter is an overview of the subject matter incorporating all the major aspects of wind turbines.

Wind Turbine

Offshore wind farm, using 5 MW turbines REpower 5M in the North Sea off the coast of Belgium.

A wind turbine is a device that converts the wind's kinetic energy into electrical power. The term appears to have been adopted from hydroelectric technology (rotary propeller). The technical description of a wind turbine is aerofoil-powered generator.

As a result of over a millennium of windmill development and modern engineering, today's wind turbines are manufactured in a wide range of vertical and horizontal axis types. The smallest turbines are used for applications such as battery charging for auxiliary power for boats or caravans or to power traffic warning signs. Slightly larger turbines can be used for making contributions to a domestic power supply while selling unused power back to the utility supplier via the electrical grid. Arrays of large turbines, known as wind farms, are becoming an increasingly important source of renewable energy and are used by many countries as part of a strategy to reduce their reliance on fossil fuels.

History

James Blyth's electricity-generating wind turbine, photographed in 1891

Windmills were used in Persia (present-day Iran) about 500-900 A.D. The windwheel of Hero of Alexandria marks one of the first known instances of wind powering a machine in history. However, the first known practical windmills were built in Sistan, an Eastern province of Iran, from the 7th century. These "Panemone" were vertical axle windmills, which had long vertical drive shafts with rectangular blades. Made of six to twelve sails covered in reed matting or cloth material, these windmills were used to grind grain or draw up water, and were used in the gristmilling and sugarcane industries.

Windmills first appeared in Europe during the Middle Ages. The first historical records of their use in England date to the 11th or 12th centuries and there are reports of German crusaders taking their windmill-making skills to Syria around 1190. By the 14th century, Dutch windmills were in use to drain areas of the Rhine delta. Advanced wind mills were described by Croatian inventor Fausto Veranzio. In his book Machinae Novae (1595) he described vertical axis wind turbines with curved or V-shaped blades.

The first electricity-generating wind turbine was a battery charging machine installed in July 1887 by Scottish academic James Blyth to light his holiday home in Marykirk, Scotland. Some months later American inventor Charles F. Brush was able to build the first automatically operated wind turbine after consulting local University professors and colleagues Jacob S. Gibbs and Brinsley Coleberd and successfully getting the blueprints peer-reviewed for electricity production in Cleveland, Ohio. Although Blyth's turbine was considered uneconomical in the United Kingdom electricity generation by wind turbines was more cost effective in countries with widely scattered populations.

The first automatically operated wind turbine, built in Cleveland in 1887 by Charles F. Brush. It was 60 feet (18 m) tall, weighed 4 tons (3.6 metric tonnes) and powered a 12 kW generator.

In Denmark by 1900, there were about 2500 windmills for mechanical loads such as pumps and mills, producing an estimated combined peak power of about 30 MW. The largest machines were on 24-meter (79 ft) towers with four-bladed 23-meter (75 ft) diameter rotors. By 1908 there were 72 wind-driven electric generators operating in the United States from 5 kW to 25 kW. Around the time of World War I, American windmill makers were producing 100,000 farm windmills each year, mostly for water-pumping.

By the 1930s, wind generators for electricity were common on farms, mostly in the United States where distribution systems had not yet been installed. In this period, high-tensile steel was cheap, and the generators were placed atop prefabricated open steel lattice towers.

A forerunner of modern horizontal-axis wind generators was in service at Yalta, USSR in 1931. This was a 100 kW generator on a 30-meter (98 ft) tower, connected to the local 6.3 kV distribution system. It was reported to have an annual capacity factor of 32 percent, not much different from current wind machines.

In the autumn of 1941, the first megawatt-class wind turbine was synchronized to a utility grid in Vermont. The Smith-Putnam wind turbine only ran for 1,100 hours before suffering a critical failure. The unit was not repaired, because of shortage of materials during the war.

The first utility grid-connected wind turbine to operate in the UK was built by John Brown & Company in 1951 in the Orkney Islands.

Despite these diverse developments, developments in fossil fuel systems almost entirely eliminated any wind turbine systems larger than supermicro size. In the early 1970s, however, anti-nuclear protests in Denmark spurred artisan mechanics to develop microturbines of 22 kW. Organizing owners into associations and co-operatives lead to the lobbying of the government and utilities and provided incentives for larger turbines throughout the 1980s and later. Local activists in Germany, nascent turbine manufacturers in Spain, and large investors in the United States in the early 1990s then lobbied for policies that stimulated the industry in those countries. Later companies formed in India and China. As of 2012, Danish company Vestas is the world's biggest wind-turbine manufacturer.

Resources

Nordex N117/2400 in Germany, a modern low-wind turbine.

Wind turbines at the Jepirachí Eolian Park in La Guajira, Colombia.

A quantitative measure of the wind energy available at any location is called the Wind Power Density (WPD). It is a calculation of the mean annual power available per square meter of swept area of a turbine, and is tabulated for different heights above ground. Calculation of wind power density includes the effect of wind velocity and air density. Color-coded maps are prepared for a particular area described, for example, as "Mean Annual Power Density at 50 Metres". In the United States, the results of the above calculation are included in an index developed by the National Renewable Energy Laboratory and referred to as "NREL CLASS". The larger the WPD, the higher it is rated by class. Classes range from Class 1 (200 watts per square meter or less at 50 m altitude) to

Class 7 (800 to 2000 watts per square m). Commercial wind farms generally are sited in Class 3 or higher areas, although isolated points in an otherwise Class 1 area may be practical to exploit.

Wind turbines are classified by the wind speed they are designed for, from class I to class IV, with A or B referring to the turbulence.

Class	Avg Wind Speed (m/s)	Turbulence
IA	10	18%
IB	10	16%
IIA	8.5	18%
IIB	8.5	16%
IIIA	7.5	18%
IIIB	7.5	16%
IVA	6	18%
IVB	6	16%

Efficiency

Not all the energy of blowing wind can be used, but some small wind turbines are designed to work at low wind speeds.

Conservation of mass requires that the amount of air entering and exiting a turbine must be equal. Accordingly, Betz's law gives the maximal achievable extraction of wind power by a wind turbine as 16/27 (59.3%) of the total kinetic energy of the air flowing through the turbine.

The maximum theoretical power output of a wind machine is thus 0.59 times the kinetic energy of the air passing through the effective disk area of the machine. If the effective area of the disk is A, and the wind velocity v, the maximum theoretical power output P is:

$$P = 0.59 \frac{1}{2} \rho v^3 A$$

where ρ is air density

As wind is free (no fuel cost), wind-to-rotor efficiency (including rotor blade friction and drag) is one of many aspects impacting the final price of wind power. Further inefficiencies, such as gearbox losses, generator and converter losses, reduce the power delivered by a wind turbine. To protect components from undue wear, extracted power is held constant above the rated operating speed as theoretical power increases at the cube of wind speed, further reducing theoretical efficiency. In 2001, commercial utility-connected turbines deliver 75% to 80% of the Betz limit of power extractable from the wind, at rated operating speed.

Efficiency can decrease slightly over time due to wear. Analysis of 3128 wind turbines older than 10 years in Denmark showed that half of the turbines had no decrease, while the other half saw a production decrease of 1.2% per year.

Types

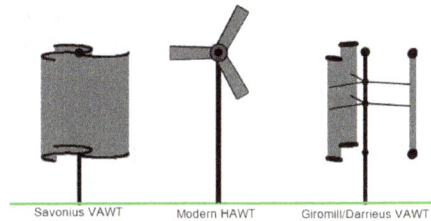

Savonius VAWT   Modern HAWT   Giromill/Darrieus VAWT

The three primary types: VAWT Savonius, HAWT towered; VAWT Darrieus as they appear in operation

Wind turbines can rotate about either a horizontal or a vertical axis, the former being both older and more common. They can also include blades (transparent or not) or be bladeless.

Horizontal Axis

Components of a horizontal axis wind turbine (gearbox, rotor shaft and brake assembly) being lifted into position

A turbine blade convoy passing through Edenfield, UK

Horizontal-axis wind turbines (HAWT) have the main rotor shaft and electrical generator at the top of a tower, and must be pointed into the wind. Small turbines are pointed by a simple wind vane, while large turbines generally use a wind sensor coupled with a servo motor. Most have a gearbox, which turns the slow rotation of the blades into a quicker rotation that is more suitable to drive an electrical generator.

Since a tower produces turbulence behind it, the turbine is usually positioned upwind of its supporting tower. Turbine blades are made stiff to prevent the blades from being pushed into the tower by high winds. Additionally, the blades are placed a considerable distance in front of the tower and are sometimes tilted forward into the wind a small amount.

Downwind machines have been built, despite the problem of turbulence (mast wake), because they don't need an additional mechanism for keeping them in line with the wind, and because in high winds the blades can be allowed to bend which reduces their swept area and thus their wind resistance. Since cyclical (that is repetitive) turbulence may lead to fatigue failures, most HAWTs are of upwind design.

Turbines used in wind farms for commercial production of electric power are usually three-bladed and pointed into the wind by computer-controlled motors. These have high tip speeds of over 320 km/h (200 mph), high efficiency, and low torque ripple, which contribute to good reliability. The blades are usually colored white for daytime visibility by aircraft and range in length from 20 to 40 meters (66 to 131 ft) or more. The tubular steel towers range from 60 to 90 meters (200 to 300 ft) tall.

The blades rotate at 10 to 22 revolutions per minute. At 22 rotations per minute the tip speed exceeds 90 meters per second (300 ft/s). A gear box is commonly used for stepping up the speed of the generator, although designs may also use direct drive of an annular generator. Some models operate at constant speed, but more energy can be collected by variable-speed turbines which use a solid-state power converter to interface to the transmission system. All turbines are equipped with protective features to avoid damage at high wind speeds, by feathering the blades into the wind which ceases their rotation, supplemented by brakes.

Year by year the size and height of turbines increase. Offshore wind turbines are built up to 8MW today and have a blade length up to 80m. Onshore wind turbines are installed in low wind speed areas and getting higher and higher towers. Usual towers of multi megawatt turbines have a height of 70 m to 120 m and in extremes up to 160 m, with blade tip speeds reaching 80 m/s to 90 m/s. Higher tip speeds means more noise and blade erosion.

Vertical Axis Design

Vertical-axis wind turbines (or VAWTs) have the main rotor shaft arranged vertically. One advantage of this arrangement is that the turbine does not need to be pointed into the wind to be effective, which is an advantage on a site where the wind direction is highly variable. It is also an advantage when the turbine is integrated into a building because it is inherently less steerable. Also, the generator and gearbox can be placed near the ground, using a direct drive from the rotor assembly to the ground-based gearbox, improving accessibility for maintenance.

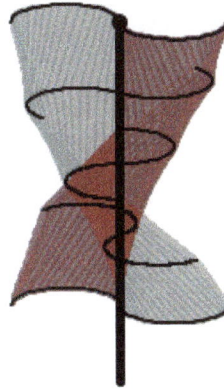

A vertical axis Twisted Savonius type turbine.

The key disadvantages include the relatively low rotational speed with the consequential higher torque and hence higher cost of the drive train, the inherently lower power coefficient, the 360-degree rotation of the aerofoil within the wind flow during each cycle and hence the highly dynamic loading on the blade, the pulsating torque generated by some rotor designs on the drive train, and the difficulty of modelling the wind flow accurately and hence the challenges of analysing and designing the rotor prior to fabricating a prototype.

When a turbine is mounted on a rooftop the building generally redirects wind over the roof and this can double the wind speed at the turbine. If the height of a rooftop mounted turbine tower is approximately 50% of the building height it is near the optimum for maximum wind energy and minimum wind turbulence. Wind speeds within the built environment are generally much lower than at exposed rural sites, noise may be a concern and an existing structure may not adequately resist the additional stress.

Subtypes of the vertical axis design include:

Offshore Horizontal Axis Wind Turbines (HAWTs) at Scroby Sands Wind Farm, UK

Onshore Horizontal Axis Wind Turbines in Zhangjiakou, China

Darrieus Wind Turbine

"Eggbeater" turbines, or Darrieus turbines, were named after the French inventor, Georges Darrieus. They have good efficiency, but produce large torque ripple and cyclical stress on the tower, which contributes to poor reliability. They also generally require some external power source, or an additional Savonius rotor to start turning, because the starting torque is very low. The torque ripple is reduced by using three or more blades which results in greater solidity of the rotor. Solidity is measured by blade area divided by the rotor area. Newer Darrieus type turbines are not held up by guy-wires but have an external superstructure connected to the top bearing.

Giromill

A subtype of Darrieus turbine with straight, as opposed to curved, blades. The cyclo-turbine variety has variable pitch to reduce the torque pulsation and is self-starting. The advantages of variable pitch are: high starting torque; a wide, relatively flat torque curve; a higher coefficient of performance; more efficient operation in turbulent winds; and a lower blade speed ratio which lowers blade bending stresses. Straight, V, or curved blades may be used.

Savonius Wind Turbine

These are drag-type devices with two (or more) scoops that are used in anemometers, *Flettner* vents (commonly seen on bus and van roofs), and in some high-reliability low-efficiency power turbines. They are always self-starting if there are at least three scoops.

Twisted Savonius

Twisted Savonius is a modified savonius, with long helical scoops to provide smooth torque. This is often used as a rooftop windturbine and has even been adapted for ships.

Another type of vertical axis is the Parallel turbine, which is similar to the crossflow fan or centrifugal fan. It uses the ground effect. Vertical axis turbines of this type have been tried for many years: a unit producing 10 kW was built by Israeli wind pioneer Bruce Brill in the 1980s.

Vortexis

The most recent advancement in Vertical Axis Wind Turbines has been the Vortexis VAWT, utilizing a pre-swirled augmented vertical axis wind turbine (PA-VAWT) designed for the purpose of developing a high efficiency VAWT concept that keeps the advantages of VAWT's compact size, lack of bias as to incoming wind direction, easy deployment and low radar cross section for use in mobile applications for the military, referred to in Special Operations as "Black Swan."

Design and Construction

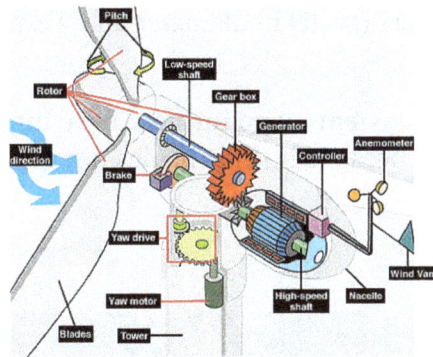

Components of a horizontal-axis wind turbine

Inside view of a wind turbine tower, showing the tendon cables.

Wind turbines are designed to exploit the wind energy that exists at a location. Aerodynamic modeling is used to determine the optimum tower height, control systems, number of blades and blade shape.

Wind turbines convert wind energy to electricity for distribution. Conventional horizontal axis turbines can be divided into three components:

- The rotor component, which is approximately 20% of the wind turbine cost, includes the blades for converting wind energy to low speed rotational energy.

- The generator component, which is approximately 34% of the wind turbine cost, includes the electrical generator, the control electronics, and most likely a gearbox (e.g. planetary gearbox), adjustable-speed drive or continuously variable transmission component for converting the low speed incoming rotation to high speed rotation suitable for generating electricity.

- The structural support component, which is approximately 15% of the wind turbine cost, includes the tower and rotor yaw mechanism.

A 1.5 MW wind turbine of a type frequently seen in the United States has a tower 80 meters (260 ft) high. The rotor assembly (blades and hub) weighs 22,000 kilograms (48,000 lb). The nacelle, which contains the generator component, weighs 52,000 kilograms (115,000 lb). The concrete base for the tower is constructed using 26,000 kilograms (58,000 lb) of reinforcing steel and contains 190 cubic meters (250 cu yd) of concrete. The base is 15 meters (50 ft) in diameter and 2.4 meters (8 ft) thick near the center.

Among all renewable energy systems wind turbines have the highest effective intensity of power-harvesting surface because turbine blades not only harvest wind power, but also concentrate it.

Unconventional Designs

The corkscrew shaped wind turbine at Progressive Field in Cleveland, Ohio

An E-66 wind turbine in the Windpark Holtriem, Germany, has an observation deck for visitors. Another turbine of the same type with an observation deck is located in Swaffham, England. Airborne wind turbine designs have been proposed and developed for many years but have yet to produce significant amounts of energy. In principle, wind turbines may also be used in conjunction with a large vertical solar updraft tower to extract the energy due to air heated by the sun.

Wind turbines which utilise the Magnus effect have been developed.

A ram air turbine (RAT) is a special kind of small turbine that is fitted to some aircraft. When deployed, the RAT is spun by the airstream going past the aircraft and can provide power for the most essential systems if there is a loss of all on-board electrical power, as in the case of the "Gimli Glider".

The two-bladed turbine SCD 6MW offshore turbine designed by aerodyn Energiesysteme, built by MingYang Wind Power has a helideck for helicopters on top of its nacelle. The prototype was erected in 2014 in Rudong China.

Turbine Monitoring and Diagnostics

Due to data transmission problems, structural health monitoring of wind turbines is usually performed using several accelerometers and strain gages attached to the nacelle to monitor the gearbox and equipments. Currently, digital image correlation and stereophotogrammetry are used to measure dynamics of wind turbine blades. These methods usually measure displacement and strain to identify location of defects. Dynamic characteristics of non-rotating wind turbines have been measured using digital image correlation and photogrammetry. Three dimensional point tracking has also been used to measure rotating dynamics of wind turbines.

Materials and Durability

Currently serving wind turbine blades are mainly made of composite materials. These blades are usually made of a polyester resin, a vinyl resin, and epoxy thermosetting matrix resin and E-glass fibers, S- glass fibers and carbon fiber reinforced materials. Construction may use manual layup techniques or composite resin injection molding. As the price of glass fibers is only about one tenth the price of carbon fiber, glass fiber is still dominant. One of the predominant ways wind turbines have gain performance is by increasing rotor diameters, and thus blade length. Longer blades place more demands on the strength and stiffness of the materials. Stiffness is especially important to avoid having blades flex to the degree that they hit the tower of the wind turbine. Carbon fiber is between 4 and 6 times stiffer than glass fiber, so carbon fiber is becoming more common in wind turbine blades.

Wind Turbines on Public Display

The Nordex N50 wind turbine and visitor centre of Lamma Winds in Hong Kong, China

A few localities have exploited the attention-getting nature of wind turbines by placing them on public display, either with visitor centers around their bases, or with viewing areas farther away. The wind turbines are generally of conventional horizontal-axis, three-bladed design, and generate power to feed electrical grids, but they also serve the unconventional roles of technology demonstration, public relations, and education.

Small Wind Turbines

A small Quietrevolution QR5 Gorlov type vertical axis wind turbine in Bristol, England. Measuring 3 m in diameter and 5 m high, it has a nameplate rating of 6.5 kW to the grid.

Small wind turbines may be used for a variety of applications including on- or off-grid residences, telecom towers, offshore platforms, rural schools and clinics, remote monitoring and other purposes that require energy where there is no electric grid, or where the grid is unstable. Small wind turbines may be as small as a fifty-watt generator for boat or caravan use. Hybrid solar and wind powered units are increasingly being used for traffic signage, particularly in rural locations, as they avoid the need to lay long cables from the nearest mains connection point. The U.S. Department of Energy's National Renewable Energy Laboratory (NREL) defines small wind turbines as those smaller than or equal to 100 kilowatts. Small units often have direct drive generators, direct current output, aeroelastic blades, lifetime bearings and use a vane to point into the wind.

Larger, more costly turbines generally have geared power trains, alternating current output, flaps and are actively pointed into the wind. Direct drive generators and aeroelastic blades for large wind turbines are being researched.

Wind Turbine Spacing

On most horizontal windturbine farms, a spacing of about 6-10 times the rotor diameter is often upheld. However, for large wind farms distances of about 15 rotor diameters should be more economically optimal, taking into account typical wind turbine and land costs. This conclusion has been reached by research conducted by Charles Meneveau of the Johns Hopkins University, and Johan Meyers of Leuven University in Belgium, based on computer simulations that take into account the detailed interactions among wind turbines (wakes) as well as with the entire turbulent atmospheric boundary layer. Moreover, recent research by John Dabiri of Caltech suggests that vertical wind turbines may be placed much more closely together so long as an alternating pattern of rotation is created allowing blades of neighbouring turbines to move in the same direction as they approach one another.

Operability

Maintenance

Wind turbines need regular maintenance to stay reliable and available, reaching 98%.

Modern turbines usually have a small onboard crane for hoisting maintenance tools and minor components. However, large heavy components like generator, gearbox, blades and so on are rarely replaced and a heavy lift external crane is needed in those cases. If the turbine has a difficult access road, a containerized crane can be lifted up by the internal crane to provide heavier lifting.

Repowering

Installation of new wind turbines can be controversial. An alternative is repowering, where existing wind turbines are replaced with bigger, more powerful ones, sometimes in smaller numbers while keeping or increasing capacity.

Demolition

Older turbines were in some early cases not required to be removed when reaching the end of their life. Some still stand, waiting to be recycled or repowered.

A demolition industry develops to recycle offshore turbines at a cost of DKK 2-4 million per MW, to be guaranteed by the owner.

Records

Fuhrländer Wind Turbine Laasow, in Brandenburg, Germany, among the world's tallest wind turbines

Éole, the largest vertical axis wind turbine, in Cap-Chat, Quebec, Canada

Largest capacity conventional drive

The Vestas V164 has a rated capacity of 8.0 MW, has an overall height of 220 m (722 ft), a diameter of 164 m (538 ft), is for offshore use, and is the world's largest-capacity wind turbine since its introduction in 2014. The conventional drive train consist of a main gearbox and a medium speed PM generator. Prototype installed in 2014 at the National Test Center Denmark nearby Østerild. Series production starts end of 2015.

Largest capacity direct drive

The Enercon E-126 with 7.58MW and 127m rotor diameter is the largest direct drive turbine. It's only for onshore use. The turbine has parted rotor blades with 2 sections for transport. In July 2016, Siemens upgraded its 7MW to 8MW.

Largest vertical-axis

Le Nordais wind farm in Cap-Chat, Quebec has a vertical axis wind turbine (VAWT) named Éole, which is the world's largest at 110 m. It has a nameplate capacity of 3.8 MW.

Largest 1 bladed turbine

Riva Calzoni M33 was a Single Bladed Wind Turbine with 350 kW, designed and built In Bologna in 1993

Largest 2 bladed turbine

Today's biggest 2 bladed turbine is build by Mingyang Wind Power in 2013. It is a SCD6.5MW offshore downwind turbine, designed by aerodyn Energiesysteme

Largest swept area

The turbine with the largest swept area is the Samsung S7.0-171, with a diameter of 171 m, giving a total sweep of 22966 m^2.

Tallest

A Nordex 3.3 MW was installed in July 2016. It has a total height of 230m, and a hub height of 164m on 100m concrete tower bottom with steel tubes on top (hybrid tower).

Vestas V164 was the tallest wind turbine, standing in Østerild, Denmark, 220 meters tall, constructed in 2014. It has a steel tube tower.

Highest tower

Fuhrländer installed a 2.5MW turbine on a 160m lattice tower in 2003

Most rotors

Lagerwey has build Four-in-One, a multi rotor wind turbine with one tower and four rotors near Maasvlakte. In April 2016, Vestas installed a 900kW quadrotor test wind turbine at Risø, made from 4 recycled 225kW V29 turbines.

Most productive

Four turbines at Rønland wind farm in Denmark share the record for the most productive wind turbines, with each having generated 63.2 GWh by June 2010.

Highest-situated

Since 2013 the world's highest-situated wind turbine was made and installed by WindAid and is located at the base of the Pastoruri Glacier in Peru at 4,877 meters (16,001 ft) above sea level. The site uses the WindAid 2.5 kW wind generator to supply power to a small rural community of micro entrepreneurs who cater to the tourists who come to the Pastoruri glacier.

Largest floating wind turbine

The world's largest—and also the first operational deep-water *large-capacity*—floating wind turbine is the 2.3 MW Hywind currently operating 10 kilometers (6.2 mi) offshore in 220-meter-deep water, southwest of Karmøy, Norway. The turbine began operating in September 2009 and utilizes a Siemens 2.3 MW turbine.

Wind Turbine Design

Wind turbine design is the process of defining the form and specifications of a wind turbine to extract energy from the wind. A wind turbine installation consists of the necessary systems needed to capture the wind's energy, point the turbine into the wind, convert mechanical rotation into electrical power, and other systems to start, stop, and control the turbine.

An example of a wind turbine, this 3 bladed turbine is the classic design of modern wind turbines

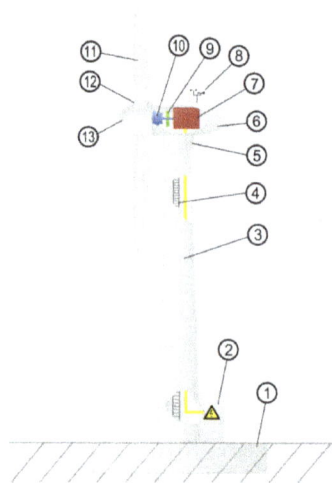

Wind turbine components : 1-Foundation, 2-Connection to the electric grid, 3-Tower, 4-Access ladder, 5-Wind orientation control (Yaw control), 6-Nacelle, 7-Generator, 8-Anemometer, 9-Electric or Mechanical Brake, 10-Gearbox, 11-Rotor blade, 12-Blade pitch control, 13-Rotor hub.

This article covers the design of horizontal axis wind turbines (HAWT) since the majority of commercial turbines use this design.

In 1919 the physicist Albert Betz showed that for a hypothetical ideal wind-energy extraction machine, the fundamental laws of conservation of mass and energy allowed no more than 16/27 (59.3%) of the kinetic energy of the wind to be captured. This Betz' law limit can be approached by modern turbine designs which may reach 70 to 80% of this theoretical limit.

In addition to aerodynamic design of the blades, design of a complete wind power system must also address design of the hub, controls, generator, supporting structure and foundation. Further design questions arise when integrating wind turbines into electrical power grids.

Aerodynamics

The shape and dimensions of the blades of the wind turbine are determined by the aerodynamic performance required to efficiently extract energy from the wind, and by the strength required to resist the forces on the blade.

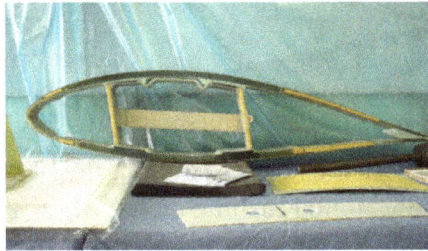

Wind rotor profile

The aerodynamics of a horizontal-axis wind turbine are not straightforward. The air flow at the blades is not the same as the airflow far away from the turbine. The very nature of the way in which energy is extracted from the air also causes air to be deflected by the turbine. In addition the aerodynamics of a wind turbine at the rotor surface exhibit phenomena that are rarely seen in other aerodynamic fields.

In 1919 the physicist Albert Betz showed that for a hypothetical ideal wind-energy extraction machine, the fundamental laws of conservation of mass and energy allowed no more than 16/27 (59.3%) of the kinetic energy of the wind to be captured. This Betz' law limit can be approached by modern turbine designs which may reach 70 to 80% of this theoretical limit.

Power Control

The speed at which a wind turbine rotates must be controlled for efficient power generation and to keep the turbine components within designed speed and torque limits. The centrifugal force on the spinning blades increases as the square of the rotation speed, which makes this structure sensitive to overspeed. Because the power of the wind increases as the cube of the wind speed, turbines have to be built to survive much higher wind loads (such as gusts of wind) than those from which they can practically generate power. Wind turbines have ways of reducing torque in high winds.

A wind turbine is designed to produce power over a range of wind speeds. All wind turbines are designed for a maximum wind speed, called the survival speed, above which they will be damaged. The survival speed of commercial wind turbines is in the range of 40 m/s (144 km/h, 89 MPH) to 72 m/s (259 km/h, 161 MPH). The most common survival speed is 60 m/s (216 km/h, 134 MPH).

If the rated wind speed is exceeded the power has to be limited. There are various ways to achieve this.

A control system involves three basic elements: sensors to measure process variables, actuators to manipulate energy capture and component loading, and control algorithms to coordinate the actuators based on information gathered by the sensors.

Stall

Stalling works by increasing the angle at which the relative wind strikes the blades (angle of attack), and it reduces the induced drag (drag associated with lift). Stalling is simple because it can be made to happen passively (it increases automatically when the winds speed up), but it increases the cross-section of the blade face-on to the wind, and thus the ordinary drag. A fully stalled turbine blade, when stopped, has the flat side of the blade facing directly into the wind.

A fixed-speed HAWT (Horizontal Axis Wind Turbine) inherently increases its angle of attack at higher wind speed as the blades speed up. A natural strategy, then, is to allow the blade to stall when the wind speed increases. This technique was successfully used on many early HAWTs. However, on some of these blade sets, it was observed that the degree of blade pitch tended to increase audible noise levels.

Vortex generators may be used to control the lift characteristics of the blade. The VGs are placed on the airfoil to enhance the lift if they are placed on the lower (flatter) surface or limit the maximum lift if placed on the upper (higher camber) surface.

Furling works by decreasing the angle of attack, which reduces the induced drag from the lift of the rotor, as well as the cross-section. One major problem in designing wind turbines is getting the blades to stall or furl quickly enough should a gust of wind cause sudden acceleration. A fully furled turbine blade, when stopped, has the edge of the blade facing into the wind.

Loads can be reduced by making a structural system softer or more flexible. This could be accomplished with downwind rotors or with curved blades that twist naturally to reduce angle of attack at higher wind speeds. These systems will be nonlinear and will couple the structure to the flow field - thus, design tools must evolve to model these nonlinearities.

Standard modern turbines all furl the blades in high winds. Since furling requires acting against the torque on the blade, it requires some form of pitch angle control, which is achieved with a slewing drive. This drive precisely angles the blade while withstanding high torque loads. In addition, many turbines use hydraulic systems. These systems are usually spring-loaded, so that if hydraulic power fails, the blades automatically furl. Other turbines use an electric servomotor for every rotor blade. They have a small battery-reserve in case of an electric-grid breakdown. Small wind turbines (under 50 kW) with variable-pitching generally use systems operated by centrifugal force, either by flyweights or geometric design, and employ no electric or hydraulic controls.

Fundamental gaps exist in pitch control, limiting the reduction of energy costs, according to a report from a coalition of researchers from universities, industry, and government, supported by the Atkinson Center for a Sustainable Future. Load reduction is currently focused on full-span blade pitch control, since individual pitch motors are the actuators currently available on commercial turbines. Significant load mitigation has been demonstrated in simulations for blades, tower, and drive train. However, there is still research needed, the methods for realization of full-span blade pitch control need to be developed in order to increase energy capture and mitigate fatigue loads.

A control technique applied to the pitch angle is done by comparing the current active power of the engine with the value of active power at the rated engine speed (active power reference, Ps reference). Control of the pitch angle in this case is done with a PI controller controls. However, in order to have a realistic response to the control system of the pitch angle, the actuator uses the time constant Tservo, an integrator and limiters so as the pitch angle to be from 0° to 30° with a rate of change (± 10° per sec).

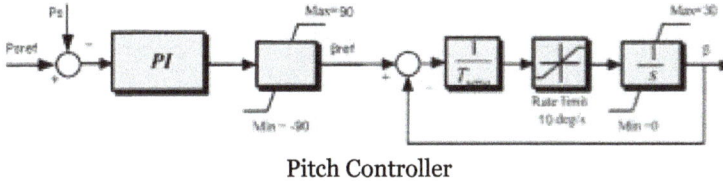

Pitch Controller

From the figure at the right, the reference pitch angle is compared with the actual pitch angle b and then the error is corrected by the actuator. The reference pitch angle, which comes from the PI controller, goes through a limiter. Restrictions on limits are very important to maintain the pitch angle in real term. Limiting the rate of change is very important especially during faults in the network. The importance is due to the fact that the controller decides how quickly it can reduce the aerodynamic energy to avoid acceleration during errors.

Other Controls

Generator Torque

Modern large wind turbines are variable-speed machines. When the wind speed is below rated, generator torque is used to control the rotor speed in order to capture as much power as possible. The most power is captured when the tip speed ratio is held constant at its optimum value (typically 6 or 7). This means that as wind speed increases, rotor speed should increase proportionally. The difference between the aerodynamic torque captured by the blades and the applied generator torque controls the rotor speed. If the generator torque is lower, the rotor accelerates, and if the generator torque is higher, the rotor slows down. Below rated wind speed, the generator torque control is active while the blade pitch is typically held at the constant angle that captures the most power, fairly flat to the wind. Above rated wind speed, the generator torque is typically held constant while the blade pitch is active.

One technique to control a permanent magnet synchronous motor is Field Oriented Control. Field Oriented Control is a closed loop strategy composed of two current controllers (an inner loop and outer loop cascade design) necessary for controlling the torque, and one speed controller.

Constant Torque Angle Control

In this control strategy the d axis current is kept zero, while the vector current is align with the q axis in order to maintain the torque angle equal with 90°. This is one of the most used control strategy because of the simplicity, by controlling only the Iqs current. So, now the electromagnetic torque equation of the permanent magnet synchronous generator is simply a linear equation depend on the Iqs current only.

So, the electromagnetic torque for Ids = 0 (we can achieve that with the d-axis controller) is now:

$$T_e = 3/2 \, p \, (\lambda_{pm} I_{qs} + (L_{ds}-L_{qs}) I_{ds} I_{qs}) = 3/2 \, p \, \lambda_{pm} I_{qs}$$

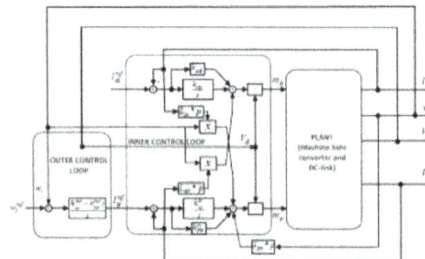

Machine Side Controller Design

So, the complete system of the machine side converter and the cascaded PI controller loops is given by the figure in the right. In that we have the control inputs, which are the duty rations m_{ds} and m_{qs}, of the PWM-regulated converter. Also, we can see the control scheme for the wind turbine in the machine side and simultaneously how we keep the I_{ds} zero (the electromagnetic torque equation is linear).

Yawing

Modern large wind turbines are typically actively controlled to face the wind direction measured by a wind vane situated on the back of the nacelle. By minimizing the yaw angle (the misalignment between wind and turbine pointing direction), the power output is maximized and non-symmetrical loads minimized. However, since the wind direction varies quickly the turbine will not strictly follow the direction and will have a small yaw angle on average. The power output losses can simply be approximated to fall with $(\cos(\text{yaw angle}))^3$. Particularly at low-to-medium wind speeds, yawing can make a significant reduction in turbine output, with wind direction variations of ±30° being quite common and long response times of the turbines to changes in wind direction. At high wind speeds, the wind direction is less variable.

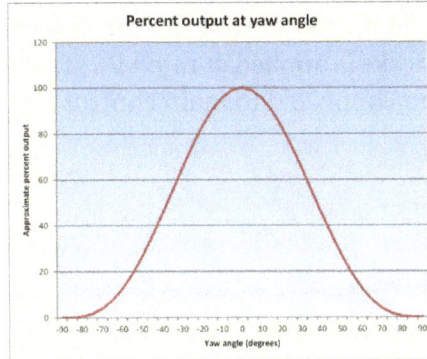

Percent output vs. wind angle

Electrical Braking

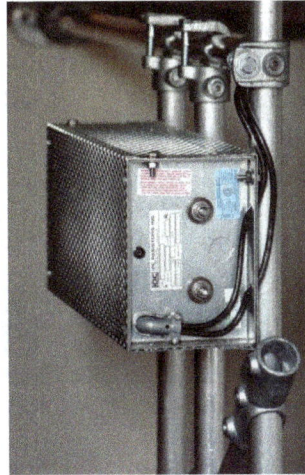

2kW Dynamic braking resistor for small wind turbine.

Braking of a small wind turbine can be done by dumping energy from the generator into a resistor bank, converting the kinetic energy of the turbine rotation into heat. This method is useful if the kinetic load on the generator is suddenly reduced or is too small to keep the turbine speed within its allowed limit.

Cyclically braking causes the blades to slow down, which increases the stalling effect, reducing the efficiency of the blades. This way, the turbine's rotation can be kept at a safe speed in faster winds while maintaining (nominal) power output. This method is usually not applied on large grid-connected wind turbines.

Mechanical Braking

A mechanical drum brake or disk brake is used to stop turbine in emergency situation such as extreme gust events or over speed. This brake is a secondary means to hold the turbine at rest for maintenance, with a rotor lock system as primary means. Such brakes are usually applied only after blade furling and electromagnetic braking have reduced the turbine speed generally 1 or 2 rotor RPM, as the mechanical brakes can create a fire inside the nacelle if used to stop the turbine from full speed. The load on the turbine increases if the brake is applied at rated RPM. Mechanical brakes are driven by hydraulic systems and are connected to main control box.

Turbine Size

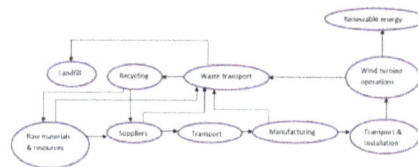

Figure 1. Flow diagram for wind turbine plant

There are different size classes of wind turbines. The smallest having power production less than 10 kW are used in homes, farms and remote applications whereas intermediate wind turbines (10-250 kW) are useful for village power, hybrid systems and distributed power. The largest wind turbines (660 kW – 2+MW) are used in central station wind farms, distributed power and community wind.

A person standing beside 15 m long blades.

For a given survivable wind speed, the mass of a turbine is approximately proportional to the cube of its blade-length. Wind power intercepted by the turbine is proportional to the square of its blade-length. The maximum blade-length of a turbine is limited by both the strength and stiffness of its material.

Labor and maintenance costs increase only gradually with increasing turbine size, so to minimize costs, wind farm turbines are basically limited by the strength of materials, and siting requirements.

Typical modern wind turbines have diameters of 40 to 90 metres (130 to 300 ft) and are rated between 500 kW and 2 MW. As of 2014 the most powerful turbine, the Vestas V-164, is rated at 8 MW and has a rotor diameter of 164m.

Nacelle

The nacelle is housing the gearbox and generator connecting the tower and rotor. Sensors detect the wind speed and direction, and motors turn the nacelle into the wind to maximize output.

Gearbox

In conventional wind turbines, the blades spin a shaft that is connected through a gearbox to the generator. The gearbox converts the turning speed of the blades 15 to 20 rotations per minute for a large, one-megawatt turbine into the faster 1,800 rotations per minute that the generator needs to generate electricity. Analysts from GlobalData estimate that gearbox market grows from $3.2bn in 2006 to $6.9bn in 2011, and to $8.1bn by 2020. Market leaders were Winergy in 2011. The use of magnetic gearboxes has also been explored as a way of reducing wind turbine maintenance costs.

Generator

Gearbox, rotor shaft and brake assembly

For large, commercial size horizontal-axis wind turbines, the electrical generator is mounted in a nacelle at the top of a tower, behind the hub of the turbine rotor. Typically wind turbines generate electricity through asynchronous machines that are directly connected with the electricity grid. Usually the rotational speed of the wind turbine is slower than the equivalent rotation speed of the electrical network: typical rotation speeds for wind generators are 5–20 rpm while a directly connected machine will have an electrical speed between 750 and 3600 rpm. Therefore, a gearbox is inserted between the rotor hub and the generator. This also reduces the generator cost and weight. Commercial size generators have a rotor carrying a field winding so that a rotating magnetic field is produced inside a set of windings called the stator. While the rotating field winding consumes a fraction of a percent of the generator output, adjustment of the field current allows good control over the generator output voltage.

Older style wind generators rotate at a constant speed, to match power line frequency, which allowed the use of less costly induction generators. Newer wind turbines often turn at whatever speed generates electricity most efficiently. The varying output frequency and voltage can be matched to the fixed values of the grid using multiple technologies such as doubly fed induction generators or full-effect converters where the variable frequency current produced is converted to DC and then back to AC. Although such alternatives require costly equipment and cause power loss, the turbine can capture a significantly larger fraction of the wind energy. In some cases, especially when turbines are sited offshore, the DC energy will be transmitted from the turbine to a central (onshore) inverter for connection to the grid.

Gearless Wind Turbine

Gearless wind turbines (also called direct drive) get rid of the gearbox completely. Instead, the rotor shaft is attached directly to the generator, which spins at the same speed as the blades. Enercon and EWT (Formerly known as Lagerwey) have produced gearless wind turbines with separately electrically excited generators for many years, and Siemens produces a gearless "inverted generator" 3 MW model while developing a 6 MW model. To make up for a direct drive generator's slower spinning rate, the diameter of the generator's rotor is increased hence containing more magnets which lets it create a lot of power when turning slowly.

Gearless wind turbines are often heavier than gear based wind turbines. A study by the EU called "Reliawind" based on the largest sample size of turbines, has shown that the reliability of gearboxes is not the main problem in wind turbines. The reliability of direct drive turbines offshore is still not known, since the sample size is so small.

Experts from Technical University of Denmark estimate that a geared generator with permanent magnets may use 25 kg/MW of the rare earth element Neodymium, while a gearless may use 250 kg/MW.

In December 2011, the US Department of Energy published a report stating critical shortage of rare earth elements such as neodymium used in large quantities for permanent magnets in gearless wind turbines. China produces more than 95% of rare earth elements, while Hitachi holds more than 600 patents covering Neodymium magnets. Direct-drive turbines require 600 kg of permanent magnet material per megawatt, which translates to several hundred kilograms of rare earth content per megawatt, as neodymium content is estimated to be 31% of magnet weight. Hybrid drivetrains (intermediate between direct drive and traditional geared) use significantly less rare earth materials. While permanent magnet wind turbines only account for about 5% of the market outside of China, their market share inside of China is estimated at 25% or higher. In 2011, demand for neodymium in wind turbines was estimated to be 1/5 of that in electric vehicles.

Blades

Blade Design

Unpainted tip of a blade

The ratio between the speed of the blade tips and the speed of the wind is called tip speed ratio. High efficiency 3-blade-turbines have tip speed/wind speed ratios of 6 to 7. Modern wind turbines are designed to spin at varying speeds. Use of aluminum and composite materials in their blades has contributed to low rotational inertia, which means that newer wind turbines can accelerate quickly if the winds pick up, keeping the tip speed ratio more nearly constant. Operating closer to their optimal tip speed ratio during energetic gusts of wind allows wind turbines to improve energy capture from sudden gusts that are typical in urban settings.

In contrast, older style wind turbines were designed with heavier steel blades, which have higher inertia, and rotated at speeds governed by the AC frequency of the power lines. The high inertia buffered the changes in rotation speed and thus made power output more stable.

It is generally understood that noise increases with higher blade tip speeds. To increase tip speed without increasing noise would allow reduction the torque into the gearbox and generator and reduce overall structural loads, thereby reducing cost. The reduction of noise is linked to the detailed aerodynamics of the blades, especially factors that reduce abrupt stalling. The inability to predict stall restricts the development of aggressive aerodynamic concepts. Some blades (mostly on Enercon) have a winglet to increase performance and/or reduce noise.

A blade can have a lift-to-drag ratio of 120, compared to 70 for a sailplane and 15 for an airliner.

The Hub

A Wind turbine hub being installed

In simple designs, the blades are directly bolted to the hub and are unable to pitch, which leads to aerodynamic stall above certain windspeeds. In other more sophisticated designs, they are bolted to the pitch mechanism, which adjusts their angle of attack according to the wind speed to control their rotational speed. The pitch mechanism is itself bolted to the hub. The hub is fixed to the rotor shaft which drives the generator directly or through a gearbox.

Blade Count

The 98 meter diameter, two-bladed NASA/DOE Mod-5B wind turbine was the largest operating wind turbine in the world in the early 1990s

The number of blades is selected for aerodynamic efficiency, component costs, and system reliability. Noise emissions are affected by the location of the blades upwind or downwind of the tower and the speed of the rotor. Given that the noise emissions from the blades' trailing edges and tips vary by the 5th power of blade speed, a small increase in tip speed can make a large difference.

The NASA test of a one-bladed wind turbine rotor configuration at Plum Brook Station near Sandusky, Ohio

Wind turbines developed over the last 50 years have almost universally used either two or three blades. However, there are patents that present designs with additional blades, such as Chan Shin's Multi-unit rotor blade system integrated wind turbine. Aerodynamic efficiency increases with number of blades but with diminishing return. Increasing the number of blades from one to two yields a six percent increase in aerodynamic efficiency, whereas increasing the blade count from two to three yields only an additional three percent in efficiency. Further increasing the blade count yields minimal improvements in aerodynamic efficiency and sacrifices too much in blade stiffness as the blades become thinner.

Theoretically, an infinite number of blades of zero width is the most efficient, operating at a high value of the tip speed ratio. But other considerations lead to a compromise of only a few blades.

Component costs that are affected by blade count are primarily for materials and manufacturing of the turbine rotor and drive train. Generally, the lower the number of blades, the lower the material and manufacturing costs will be. In addition, the lower the number of blades, the higher the rotational speed can be. This is because blade stiffness requirements to avoid interference with the tower limit how thin the blades can be manufactured, but only for upwind machines; deflection of blades in a downwind machine results in increased tower clearance. Fewer blades with higher rotational speeds reduce peak torques in the drive train, resulting in lower gearbox and generator costs.

System reliability is affected by blade count primarily through the dynamic loading of the rotor into the drive train and tower systems. While aligning the wind turbine to changes in wind direction (yawing), each blade experiences a cyclic load at its root end depending on blade position. This is true of one, two, three blades or more. However, these cyclic loads when combined together at the drive train shaft are symmetrically balanced for three blades, yielding smoother operation during turbine yaw. Turbines with one or two blades can use a pivoting teetered hub to also nearly eliminate the cyclic loads into the drive shaft and system during yawing. A Chinese 3.6 MW two-blade is being tested in Denmark. Mingyang won a bid for 87 MW (29 * 3 MW) two-bladed offshore wind turbines near Zhuhai in 2013.

Finally, aesthetics can be considered a factor in that some people find that the three-bladed rotor is more pleasing to look at than a one- or two-bladed rotor.

Blade Materials

Several modern wind turbines use rotor blades with carbon-fibre girders to reduce weight.

In general, ideal materials should meet the following criteria:

- wide availability and easy processing to reduce cost and maintenance

- low weight or density to reduce gravitational forces

- high strength to withstand strong loading of wind and gravitational force of the blade itself

- high fatigue resistance to withstand cyclic loading

- high stiffness to ensure stability of the optimal shape and orientation of the blade and clearance with the tower

- high fracture toughness

- the ability to withstand environmental impacts such as lightning strikes, humidity, and temperature

Wood and canvas sails were used on early windmills due to their low price, availability, and ease of manufacture. Smaller blades can be made from light metals such as aluminium. These materials, however, require frequent maintenance. Wood and canvas construction limits the airfoil shape to a flat plate, which has a relatively high ratio of drag to force captured (low aerodynamic efficiency) compared to solid airfoils. Construction of solid airfoil designs requires inflexible materials such as metals or composites. Some blades also have incorporated lightning conductors.

New wind turbine designs push power generation from the single megawatt range to upwards of 10 megawatts using larger and larger blades. A larger area effectively increases the tip-speed ratio of a turbine at a given wind speed, thus increasing its energy extraction. Computer-aided engineering software such as HyperSizer (originally developed for spacecraft design) can be used to improve blade design.

As of 2015 the rotor diameters of onshore wind turbine blades are as large as 130 meters, while the diameter of offshore turbines reach 170 meters. In 2001, an estimated 50 million kilograms of fibreglass laminate were used in wind turbine blades.

An important goal of larger blade systems is to control blade weight. Since blade mass scales as the cube of the turbine radius, loading due to gravity constrains systems with larger blades. Gravitational loads include axial and tensile/ compressive loads (top/bottom of rotation) as well as bending (lateral positions). The magnitude of these loads fluctuates cyclically and the edgewise moments are reversed every $180°$ of rotation. Typical rotor speeds and design life are ~10rpm and 20 years, respectively, with the number of lifetime revolutions on the order of 10^8. Considering wind, it is expected that turbine blades go through ~10^9 loading cycles. Wind is another source of rotor blade loading. Lift causes bending in the flapwise direction (out of rotor plane) while air flow around the blade cause edgewise bending (in the rotor plane). Flapwise bending involves tension on the pressure (upwind) side and compression on the suction (downwind) side. Edgewise bending involves tension on the leading edge and compression on the trailing edge.

Wind loads are cyclical because of natural variability in wind speed and wind shear (higher speeds at top of rotation).

Failure in ultimate loading of wind-turbine rotor blades exposed to wind and gravity loading is a failure mode that needs to be considered when the rotor blades are designed. The wind speed that causes bending of the rotor blades exhibits a natural variability, and so does the stress response in the rotor blades. Also, the resistance of the rotor blades, in terms of their tensile strengths, exhibits a natural variability.

In light of these failure modes and increasingly larger blade systems, there has been continuous effort toward developing cost-effective materials with higher strength-to-mass ratios. In order to extend the current 20 year lifetime of blades and enable larger area blades to be cost-effective, the design and materials need to be optimized for stiffness, strength, and fatigue resistance.

The majority of current commercialized wind turbine blades are made from fiber-reinforced polymers (FRP's), which are composites consisting of a polymer matrix and fibers. The long fibers provide longitudinal stiffness and strength, and the matrix provides fracture toughness, delamination strength, out-of-plane strength, and stiffness. Material indices based on maximizing power efficiency, and having high fracture toughness, fatigue resistance, and thermal stability, have been shown to be highest for glass and carbon fiber reinforced plastics (GFRP's and CFRPs).

Fiberglass-reinforced epoxy blades of Siemens SWT-2.3-101 wind turbines. The blade size of 49 meters is in comparison to a substation behind them at Wolfe Island Wind Farm.

Manufacturing blades in the 40 to 50 metre range involves proven fibreglass composite fabrication techniques. Manufactures such as Nordex SE and GE Wind use an infusion process. Other manufacturers use variations on this technique, some including carbon and wood with fibreglass in an epoxy matrix. Other options include preimpregnated ("prepreg") fibreglass and vacuum-assisted resin transfer molding. Each of these options use a glass-fibre reinforced polymer composite constructed with differing complexity. Perhaps the largest issue with more simplistic, open-mould, wet systems are the emissions associated with the volatile organics released. Preimpregnated materials and resin infusion techniques avoid the release of volatiles by containing all VOC's. However, these contained processes have their own challenges, namely the production of thick laminates necessary for structural components becomes more difficult. As the preform resin permeability dictates the maximum laminate thickness, bleeding is required to eliminate voids and ensure proper resin distribution. One solution to resin distribution a partially preimpregnated fibreglass. During evacuation, the dry fabric provides a path for airflow and, once heat and pressure are applied, resin may flow into the dry region resulting in a thoroughly impregnated laminate structure.

Epoxy-based composites have environmental, production, and cost advantages over other resin systems. Epoxies also allow shorter cure cycles, increased durability, and improved surface finish. Prepreg operations further reduce processing time over wet lay-up systems. As turbine blades pass 60 metres, infusion techniques become more prevalent; the traditional resin transfer moulding injection time is too long as compared to the resin set-up time, limiting laminate thickness. Injection forces resin through a thicker ply stack, thus depositing the resin where in the laminate structure before gelation occurs. Specialized epoxy resins have been developed to customize lifetimes and viscosity.

Carbon fibre-reinforced load-bearing spars can reduce weight and increase stiffness. Using carbon fibres in 60 metre turbine blades is estimated to reduce total blade mass by 38% and decrease cost by 14% compared to 100% fibreglass. Carbon fibres have the added benefit of reducing the thickness of fiberglass laminate sections, further addressing the problems associated with resin wetting of thick lay-up sections. Wind turbines may also benefit from the general trend of increasing use and decreasing cost of carbon fibre materials.

Although glass and carbon fibers have many optimal qualities for turbine blade performance, there are several downsides to these current fillers, including the fact that high filler fraction (10-70 wt%) causes increased density as well as microscopic defects and voids that often lead to premature failure.

Recent developments include interest in using carbon nanotubes (CNT's) to reinforce polymer-based nanocomposites. CNT's can be grown or deposited on the fibers, or added into polymer resins as a matrix for FRP structures. Using nanoscale CNT's as filler instead of traditional microscale filler (such as glass or carbon fibers) results in CNT/polymer nanocomposites, for which the properties can be changed significantly at very low filler contents (typically < 5 wt%). They have very low density, and improve the elastic modulus, strength, and fracture toughness of the polymer matrix. The addition of CNT's to the matrix also reduces the propagation of interlaminar cracks which can be a problem in traditional FRP's.

Further improvement is possible through the use of carbon nanofibers (CNF's) in the blade coatings. A major problem in desert environments is erosion of the leading edges of blades by wind carrying sand, which increases roughness and decreases aerodynamic performance. The particle erosion resistance of fiber-reinforced polymers is poor when compared to metallic materials and elastomers, and needs to be improved. It has been shown that the replacement of glass fiber with CNF on the composite surface greatly improves erosion resistance. CNF's have also been shown to provide good electrical conductivity (important for lightning strikes), high damping ratio, and good impact-friction resistance. These properties make CNF-based nanopaper a prospective coating for wind turbine blades.

Tower

Tower Height

Wind velocities increase at higher altitudes due to surface aerodynamic drag (by land or water surfaces) and the viscosity of the air. The variation in velocity with altitude, called wind shear, is most dramatic near the surface. Typically, the variation follows the wind profile power law, which predicts that wind speed rises proportionally to the seventh root of altitude. Doubling the altitude of a turbine, then, increases the expected wind speeds by 10% and the expected power by 34%. To avoid buckling, doubling the tower height generally requires doubling the diameter of the tower as well, increasing the amount of material by a factor of at least four.

At night time, or when the atmosphere becomes stable, wind speed close to the ground usually subsides whereas at turbine hub altitude it does not decrease that much or may even increase. As a result, the wind speed is higher and a turbine will produce more power than expected from the 1/7 power law: doubling the altitude may increase wind speed by 20% to 60%. A stable atmosphere is caused by radiative cooling of the surface and is common in a temperate climate: it usually occurs when there is a (partly) clear sky at night. When the (high altitude) wind is strong (a 10-meter wind speed higher than approximately 6 to 7 m/s) the stable atmosphere is disrupted because of friction turbulence and the atmosphere will turn neutral. A daytime atmosphere is either neutral (no net radiation; usually with strong winds and heavy clouding) or unstable (rising air because of ground heating—by the sun). Here again the 1/7 power law applies or is at least a good approximation of the wind profile. Indiana had been rated as having a wind capacity of 30,000 MW, but by raising the expected turbine height from 50 m to 70 m, the wind capacity estimate was raised to 40,000 MW, and could be double that at 100 m.

For HAWTs, tower heights approximately two to three times the blade length have been found to balance material costs of the tower against better utilisation of the more expensive active components.

Sections of a wind turbine tower, transported in a bulk carrier ship

Road size restrictions makes transportation of towers with a diameter of more than 4.3 m difficult. Swedish analyses show that it is important to have the bottom wing tip at least 30 m above the tree tops, but a taller tower requires a larger tower diameter. A 3 MW turbine may increase output from 5,000 MWh to 7,700 MWh per year by going from 80 to 125 meter tower height. A tower profile made of connected shells rather than cylinders can have a larger diameter and still be transportable. A 100 m prototype tower with TC bolted 18 mm 'plank' shells has been erected at the wind turbine test center Høvsøre in Denmark and certified by Det Norske Veritas, with a Siemens nacelle. Shell elements can be shipped in standard 12 m shipping containers, and 2½ towers per week are produced this way.

As of 2003, typical modern wind turbine installations use towers about 210 ft (65 m) high. Height is typically limited by the availability of cranes. This has led to a variety of proposals for "partially self-erecting wind turbines" that, for a given available crane, allow taller towers that put a turbine in stronger and steadier winds, and "self-erecting wind turbines" that can be installed without cranes.

Tower Materials

Currently, the majority of wind turbines are supported by conical tubular steel towers. These towers represent 30% – 65% of the turbine weight and therefore account for a large percentage of the turbine transportation costs. The use of lighter materials in the tower could greatly reduce the overall transport and construction cost of wind turbines, however the stability must be maintained. Higher grade S500 steel costs 20%-25% more than S335 steel (standard structural steel), but it requires 30% less material because of its improved strength. Therefore, replacing wind turbine towers with S500 steel would result in a net savings in both weight and cost.

Another disadvantage of conical steel towers is that constructing towers that meet the requirements of wind turbines taller than 90 meters proves challenging. High performance concrete shows potential to increase tower height and increase the lifetime of the towers. A hybrid of prestressed concrete and steel has shown improved performance over standard tubular steel at tower heights of 120 meters. Concrete also gives the benefit of allowing for small precast sections to be assembled on site, avoiding the challenges steel faces during transportation. One downside of concrete towers is the higher CO_2 emissions during concrete production as compared to steel. However, the overall environmental benefit should be higher if concrete towers can double the wind turbine lifetime.

Wood is being investigated as a material for wind turbine towers, and a 100 metre tall tower supporting a 1.5 MW turbine has been erected in Germany. The wood tower shares the same transportation benefits of the segmented steel shell tower, but without the steel resource consumption.

Connection to the Electric Grid

All grid-connected wind turbines, from the first one in 1939 until the development of variable-speed grid-connected wind turbines in the 1970s, were fixed-speed wind turbines. As recently as 2003, nearly all grid-connected wind turbines operated at exactly constant speed (synchronous generators) or within a few percent of constant speed (induction generators). As of 2011, many operational wind turbines used fixed speed induction generators (FSIG). As of 2011, most new grid-connected wind turbines are variable speed wind turbines—they are in some variable speed configuration.

Early wind turbine control systems were designed for peak power extraction, also called maximum power point tracking—they attempt to pull the maximum possible electrical power from a given wind turbine under the current wind conditions. More recent wind turbine control systems deliberately pull less electrical power than they possibly could in most circumstances, in order to provide other benefits, which include:

- spinning reserves to quickly produce more power when needed—such as when some other generator suddenly drops from the grid—up to the max power supported by the current wind conditions.

- Variable-speed wind turbines can (very briefly) produce more power than the current wind conditions can support, by storing some wind energy as kinetic energy (accelerating during brief gusts of faster wind) and later converting that kinetic energy to electric energy (decelerating, either when more power is needed elsewhere, or during short lulls in the wind, or both).

- damping (electrical) subsynchronous resonances in the grid

- damping (mechanical) resonances in the tower

The generator in a wind turbine produces alternating current (AC) electricity. Some turbines drive an AC/AC converter—which converts the AC to direct current (DC) with a rectifier and then back to AC with an inverter—in order to match the frequency and phase of the grid. However, the most common method in large modern turbines is to instead use a doubly fed induction generator directly connected to the electricity grid.

A useful technique to connect a permanent magnet synchronous generator to the grid is by using a back-to-back converter. Also, we can have control schemes so as to achieve unity power factor in the connection to the grid. In that way the wind turbine will not consume reactive power, which is the most common problem with wind turbines that use induction machines. This leads to a more stable power system. Moreover, with different control schemes a wind turbine with a permanent magnet synchronous generator can provide or consume reactive power. So, it can work as a dynamic capacitor/inductor bank so as to help with the power systems' stability.

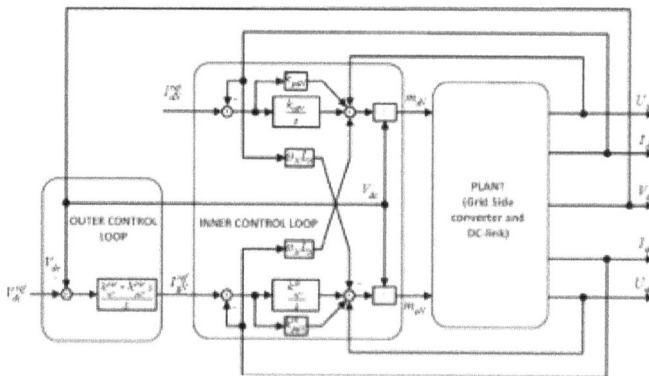

Grid Side Controller Design

Below we show the control scheme so as to achieve unity power factor :

Reactive power regulation consists of one PI controller in order to achieve operation with unity power factor (i.e. $Q_{grid} = 0$). It is obvious that I_{dN} has to be regulated to reach zero at steady-state ($I_{dNref} = 0$).

We can see the complete system of the grid side converter and the cascaded PI controller loops in the figure in the right.

Foundations

Wind turbine foundations

Wind turbines, by their nature, are very tall slender structures, this can cause a number of issues when the structural design of the foundations are considered.

The foundations for a conventional engineering structure are designed mainly to transfer the vertical load (dead weight) to the ground, this generally allows for a comparatively unsophisticated arrangement to be used. However, in the case of wind turbines, due to the high wind and environmental loads experienced there is a significant horizontal dynamic load that needs to be appropriately restrained.

This loading regime causes large moment loads to be applied to the foundations of a wind turbine. As a result, considerable attention needs to be given when designing the footings to ensure that the turbines are sufficiently restrained to operate efficiently. In the current Det Norske Veritas (DNV) guidelines for the design of wind turbines the angular deflection of the foundations are limited to 0.5°. DNV guidelines regarding earthquakes suggest that horizontal loads are larger than vertical loads for offshore wind turbines, while guidelines for tsunamis only suggest designing for maximum sea waves. In contrast, IEC suggests considering tsunami loads.

Scale model tests using a 50-g centrifuge are being performed at the Technical University of Denmark to test monopile foundations for offshore wind turbines at 30 to 50-m water depth.

Costs

Liftra *Blade Dragon* installing a single blade on wind turbine hub.

The modern wind turbine is a complex and integrated system. Structural elements comprise the majority of the weight and cost. All parts of the structure must be inexpensive, lightweight, durable, and manufacturable, under variable loading and environmental conditions. Turbine systems that have fewer failures, require less maintenance, are lighter and last longer will lead to reducing the cost of wind energy.

One way to achieve this is to implement well-documented, validated analysis codes, according to a 2011 report from a coalition of researchers from universities, industry, and government, supported by the Atkinson Center for a Sustainable Future.

The major parts of a modern turbine may cost (percentage of total): tower 22%, blades 18%, gearbox 14%, generator 8%.

Efficiency and Wind Speed

The efficiency of a wind turbine is maximum at its design wind velocity, and efficiency decreases with the fluctuations in wind. The lowest velocity at which the turbine develops its full power is known as rated wind velocity. Below some minimum wind velocity, no useful power output can be produced from wind turbine. There are limits on both the minimum (2-5 m/s) and maximum (25-30 m/s) wind velocity for the efficient operation of wind turbines.

Conservation of mass requires that the amount of air entering and exiting a turbine must be equal. Accordingly, Betz's law gives the maximal achievable extraction of wind power by a wind turbine as 16/27 (59.3%) of the total kinetic energy of the air flowing through the turbine.

The maximum theoretical power output of a wind machine is thus 0.59 times the kinetic energy of the air passing through the effective disk area of the machine. If the effective area of the disk is A, and the wind velocity v, the maximum theoretical power output P is:

where ρ is air density

As wind is free (no fuel cost), wind-to-rotor efficiency (including rotor blade friction and drag) is one of many aspects impacting the final price of wind power. Further inefficiencies, such as gearbox losses, generator and converter losses, reduce the power delivered by a wind turbine. To protect components from undue wear, extracted power is held constant above the rated operating speed as theoretical power increases at the cube of wind speed, further reducing theoretical efficiency. In 2001, commercial utility-connected turbines deliver 75% to 80% of the Betz limit of power extractable from the wind, at rated operating speed.

All power plants have some consumption when they produce power, and some standby consumption when they are turned on without producing power. For a modern 3 MW wind turbine, the consumption may be 6-58 kW depending on circumstances.

Design Specification

The design specification for a wind-turbine will contain a power curve and guaranteed availability. With the data from the wind resource assessment it is possible to calculate commercial viability. The typical operating temperature range is −20 to 40 °C (−4 to 104 °F). In areas with extreme climate (like Inner Mongolia or Rajasthan) specific cold and hot weather versions are required.

Wind turbines can be designed and validated according to IEC 61400 standards.

Low Temperature

Utility-scale wind turbine generators have minimum temperature operating limits which apply in areas that experience temperatures below −20 °C. Wind turbines must be protected from ice accumulation. It can make anemometer readings inaccurate and which, in certain turbine control designs, can cause high structure loads and damage. Some turbine manufacturers offer low-temperature packages at a few percent extra cost, which include internal heaters, different lubricants, and different alloys for structural elements. If the low-temperature interval is combined with a low-wind condition, the wind turbine will require an external supply of power, equivalent to a few percent of its rated power, for internal heating. For example, the St. Leon, Manitoba project has a total rating of 99 MW and is estimated to need up to 3 MW (around 3% of capacity) of station service power a few days a year for temperatures down to −30 °C. This factor affects the economics of wind turbine operation in cold climates.

Airborne Wind Turbine

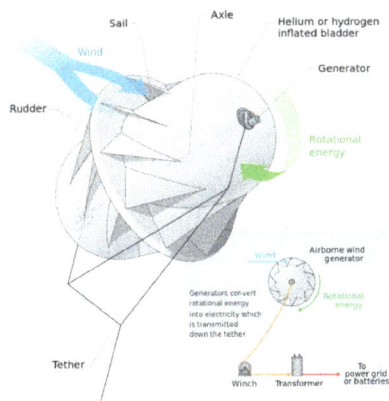

Airborne wind generator of flip-wing style

An airborne wind turbine is a design concept for a wind turbine with a rotor supported in the air without a tower, thus benefiting from more mechanical and aerodynamic options, the higher velocity and persistence of wind at high altitudes, while avoiding the expense of tower construction, or the need for slip rings or yaw mechanism. An electrical generator may be on the ground or airborne. Challenges include safely suspending and maintaining turbines hundreds of meters off the ground in high winds and storms, transferring the harvested and/or generated power back to earth, and interference with aviation.

Airborne wind turbines may operate in low or high altitudes; they are part of a wider class of Airborne Wind Energy Systems (AWES) addressed by high-altitude wind power and crosswind kite power. When the generator is on the ground, then the tethered aircraft need not carry the generator mass or have a conductive tether. When the generator is aloft, then a conductive tether would be used to transmit energy to the ground or used aloft or beamed to receivers using microwave or laser. Kites and 'helicopters' come down when there is insufficient wind; kytoons and blimps may resolve the matter with other disadvantages. Also, bad weather such as lightning or thunderstorms, could temporarily suspend use of the machines, probably requiring them to be brought back down to the ground and covered. Some schemes require a long power cable and, if the turbine is high enough, a prohibited airspace zone. As of July 2015, no commercial airborne wind turbines are in regular operation.

Aerodynamic Variety

An aerodynamic airborne wind power system relies on the wind for support.

Crosswind kite generator with fast motion transfer

Miles L. Loyd proposed and analyzed an efficient AWES in his work "Crosswind Kite Power" in 1980. Power output of AWES with crosswind wing motion is proportional to a square of a lift/drag ratio of the wing. Such AWES is based on the same aerodynamic principles as a conventional wind turbine (AWES), but it is more efficient because the air speed is constant along the wing span and the aerodynamic forces are resisted by tension of a tether, rather than by bending of a tower.

Bryan Roberts, a professor of engineering at the University of Technology, in Sydney, Australia, has proposed a helicopter-like craft which flies to 15,000 feet (4,600 m) altitude and stays there, held aloft by wings that generate lift from the wind, and held in place by a cable to a ground anchor. According to its designers, while some of the energy in the wind would be 'lost' on lift, the constant and potent winds would allow it to generate constant electricity. Since the winds usually blow horizontally, the turbines would be at an angle from the horizontal, catching winds while still generating lift. Deployment could be done by feeding electricity to the turbines, which would turn them into electric motors, lifting the structure into the sky.

The Dutch ex-astronaut and physicist Wubbo Ockels, working with the Delft University of Technology in the Netherlands, has designed and demonstrated an airborne wind turbine he called a "Laddermill". It consists of an endless loop of kites. The kites lift up one end of the endless loop (the "ladder"), and the released energy is used to drive an electric generator.

A Sept'09 paper from Carbon Tracking Ltd., Ireland has shown the capacity factor of a kite using ground-based generation to be 52.2%, which compares favorably with terrestrial wind-farm capacity factors of 30%.

A team from Worcester Polytechnic Institute in the United States has developed a smaller-scale with an estimated output of about 1 kW. It uses a kiteboarding kite to induce a rocking motion in a pivoting beam.

The Kitegen uses a prototype vertical-axis wind turbine. It is an innovative plan (still in the construction phase) that consists of one wind farm with a vertical spin axis, and employs kites to exploit high-altitude winds. The Kite Wind Generator (KWG) or Kitegen is claimed to eliminate all the static and dynamic problems that prevent the increase of the power (in terms of dimensions) obtainable from the traditional horizontal-axis wind turbine generators. Generating equipment would remain on the ground, and only the airfoils are supported by the wind. Such a wind power plant would be capable of producing the energy equivalent to a nuclear plant, while using an area of few square kilometres, without occupying it exclusively. (The majority of this area can still be used for agriculture, or navigation in the case of an offshore installation.)

The Rotokite is developed from Gianni Vergnano's idea. It uses aerodynamic profiles similar to kites that have been rotated on their own axis, emulating the performance of a propeller. The use of the rotation principle simplifies the problem of checking the flight of the kites and eliminates the difficulties due to the lengths of cables, enabling the production of wind energy at low cost. The Heli Wind Power is a project of Gianni Vergnano that uses a tethered kite.

The HAWE System is developed from Tiago Pardal's idea. The System that consists in a Pumping Cycle similar to kite systems. In Generation Phase the pulling force increase 5-10 times due to Magnus Effect of a spinning cylinder(aerial platform), like a kite the pulling force produced by the aerial platform will unwind the cable and generate electricity in the ground.In the Recovery Phase it rewinds the cable with no Magnus Effect in the aerial platform.

In August 2011 the German company SkySails, producer of kites for ship propulsion, announced a kite-based wind power system for on- and offshore applications that is supposed to be "30% cheaper than current offshore solutions".

In June 2012, the German company NTS GmbH had successfully tested X-Wind technology (spoken: Cross-Wind) on linear rail system in Freidland, Germany. "NTS Energie- und Transportsysteme GmbH" was found in 2006 by Uwe Ahrens. X-Wind technology combines two well-known technologies - automatically steered kites and generators on a closed loop rail system. Closed loop prototype is under construction at Mecklenburg-Vorpommern, Germany. This technology allows to harness increasingly stable and constant wind currents at altitudes between 200 and 500m. Technical report readings and measurements show that NTS X-Wind Systems double to triple the efficiency of conventional wind energy systems according to energy production.

In May 2013, the Californian company Makani Power, developer of some crosswind hybrid kite systems with onboard generator doubling as motor, has been acquired by Google.

In May 2013, an airborne wind energy system with a ground-based generator using fast motion transfer was suggested by L. Goldstein.

In 2015, a sails on rope wind and ocean current energy system was invented by a Taiwanese Johnson Hsu.

On 15 December 2015 Windswept and Interesting Ltd demonstrated a "Daisy" kite ring stack airborne wind turbine. The Daisy kite stack demonstrated on 15 December 2015 is the only airborne wind energy system to have won the someawe.org 100*3 AWE challenge. The Daisy system uses tensioned torsion transfer of kite motion to turn a ground based generator.

Aerostat Variety

An aerostat-type wind power system relies at least in part on buoyancy to support the wind-collecting elements. Aerostats vary in their designs and resulting lift-over-drag aerodynamic characteristic; the kiting effect of higher lift-over-drag shapes for the aerostat can effectively keep an airborne turbine aloft; a variety of such kiting balloons were made famous in the kytoon by Domina C. Jalbert.

Balloons can be incorporated to keep systems up without wind, but balloons leak slowly and have to be resupplied with lifting gas, possibly patched as well. Very large, sun heated balloons may solve the helium or hydrogen leakage problems.

An Ontario based company called Magenn is developing a turbine called the Magenn Air Rotor System (MARS). A future 1,000-foot (300 m)-wide MARS system would use a horizontal rotor in a helium suspended apparatus which is tethered to a transformer on the ground. Magenn claims that their technology provides high torque, low starting speeds, and superior overall efficiency thanks to its ability to deploy higher in comparison to non-aerial solutions. The first prototypes were built by TCOM in April 2008. No production units have been delivered.

Boston-based Altaeros Energies uses a helium-filled balloon shroud to lift a wind turbine into the air, transferring the resultant power down to a base station through the same cables used to control the shroud. A 35-foot prototype using a standard Skystream 2.5kW 3.7m wind turbine was flown and tested in 2012. In fall 2013, Altaeros was at work on its first commercial-scale demonstration in Alaska.

Concept drawing of the Twind technology.

The Twind Technology concept uses a pair of captive balloons at an altitude of 800 meters. The tether cables transmit force to a rotating platform on the ground. Each balloon has a sail connected to it. The two balloons move alternately, the balloon with the sail open moves downwind and draws the other balloon upwind, and then the motion reverses. The tether cable can be used to turn the shaft of a generator to produce electrical energy or perform other works (grinding, sawing, pumping).

Estimated Costs

Peer reviewed papers show that certain types of AWES can produce energy ten times cheaper than conventional wind turbines, which is less than $0.02 per kWh.

Crosswind Kite Power

Crosswind kite power is power derived from a class of airborne wind-energy conversion systems (AWECS, aka AWES) or crosswind kite power systems (CWKPS) characterized by a kite system that has energy-harvesting parts that fly transverse to the direction of the ambient wind, i.e., to crosswind mode; sometimes the entire wing set and tether set is flown in crosswind mode. These systems at many scales from toy to power-grid-feeding sizes may be used as high-altitude wind power (HAWP) devices or low-altitude wind power (LAWP) devices without having to use towers. Flexible wings or rigid wings may be used in the kite system. A tethered wing, flying in crosswind at many times wind speed, harvests wind power from an area that is many times exceeding the wing's own area. Crosswind kite power systems have some advantages over conventional wind turbines: access to more powerful and stable wind resource, high capacity factor, capability for deployment on and offshore at comparable costs, and no need for a tower. Additionally, the wings of the CWKPS may vary in aerodynamic efficiency; the movement of crosswinding tethered wings is sometimes compared with the outer parts of conventional wind turbine blades. However, a conventional traverse-to-wind rotating blade set carried aloft in a kite-power system has the blade set cutting to crosswind and is a form of crosswind kite power. Miles L. Loyd furthered studies on crosswind kite power systems in his work "Crosswind Kite Power" in 1980. Some believe that crosswind kite power was introduced by P. Payne and C. McCutchen in their patent No. 3,987,987, filed in 1975, however, crosswind kite power was used far before such patent, e.g., in target kites for war-target practice where the crosswinding power permitted high speeds to give practice to gunners.

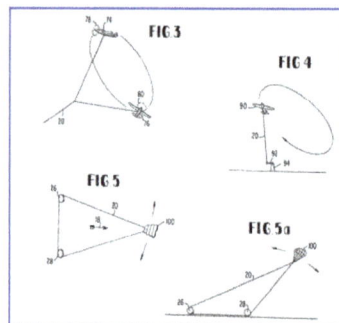

US3987987figs

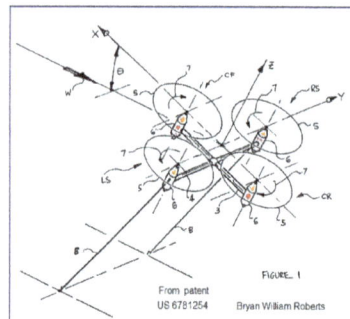

This illustrates where parts of the wing set of a crosswind kite power device is crosswinding during conversion of the wind's kinetic energy.

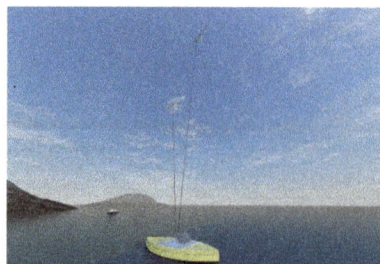

Crosswind kite power station with separate motion transfer with two wings offshore, artist's impression.

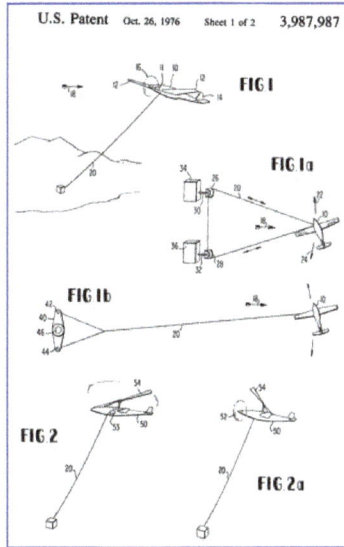

Drawing from patent US 3,987,987.

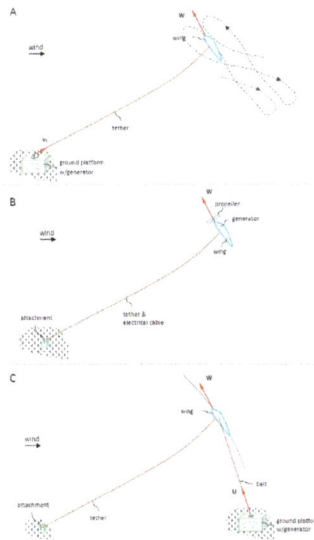

Scheme of some types of airborne crosswind power systems

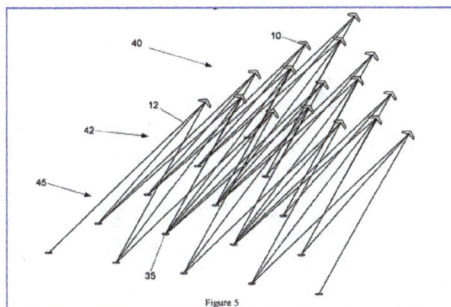

BenjaminTignerFig5US8066225 Farming crosswind kite power devices is illustrated by Benjamin Tigner.

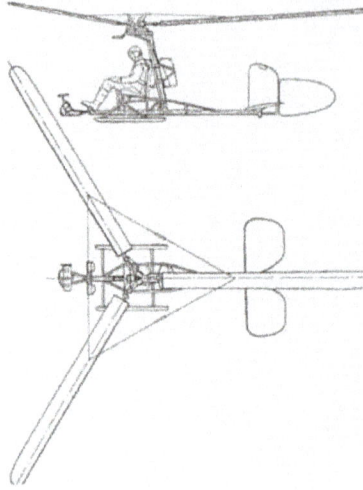

Lifter of human by use of fast-motion autorotating bladed crosswind kite power system, a gyrokite type.

Charvolants2 As George Pocock control his kite system wings to left or right, he would gain power because of the crosswind kite power energy-gain effect. He has followers considering him the father of traction kiting using crosswind kite power in his transport operations.

Kite-Wahnsinn am Silvaplana See. The crosswind kite power systems shown has a purpose to move the athlete fast downwind, upwind, and sometimes into the air to significant altitudes and distances. The wind is slightly slowed by such activity, as the CWKPS harvests energy from the wind. Kiteboarding.

Types of Crosswind Kite Power Systems (CWKPS)

How a system extracts energy from the wind and transfers energy to useful purposes helps to define types of crosswind kite power systems. One typing parameter regards the position of the generator or pump or tasking line or device. Another typing parameter regards how the tethers of the tether set of the kite system are utilized; the tethers holding the kiting wing elements aloft may be used in various ways to form types; tethers may simply hold working wings aloft, or they may be pulling loads on the ground, or multitasking by sending aloft-gained electricity to ground receivers or by pulling loads or by being the tasking device itself as when used for pulling people or things or cutting or grinding things. Some types are distinguished by fast motion transfer or slow motion transfer. Typing of crosswind kite power system also occurs by the nature of the wing set where count of wings and types of wings matter to designers and users; a wing set might be in a train arrangement, stack configuration, arch complex, dome mesh, coordinating family of wings, or just be a simple single-wing with single tether. Types of crosswind kite power devices are also distinguished by scale, purpose, intended life, and cost level. Typing by economic success occurs; is the system effective in the energy or task market or not? Some CWKPS are a type called lifters; they are purposed just for lifting loads, perhaps humans; the type is frequented by the use of autorotating blades that appear then to look like helicopters. A single crosswind kite power system (CWKPS) may be a hybrid complex performing aloft energy generation while also performing ground-based work through tether pulling of loads. The crosswind kite power systems that involve fluttering elements are being explored in several research centers; flutter is mined for energy conversion in a few ways. Researchers are showing types of CWKPS that are difficult to classify or type.

Tether Pulling of People or Goods on Boards, in Hulls, with Skis, etc.

In the systems of this type of CWKPS, the pulling tether set drives the resisting people and objects to various points on the surface of water bodies or land or points in the atmosphere. In this type of crosswind kite power operation, the design of the resistive objects (people, boards, hulls, boats, ships, water turbines, air turbines, other wings) makes for further types. Crosswinding of the upper flying wings provide power to achieve certain final objectives. The objectives are found in such as kiteboarding, kite windsurfing, snowkiting, yacht kiting, freighter-ship sailing, kite boating, and free-flight soaring and jumping. A collection of researchers have explored the historic free-flight parakite realm to where crosswind flying of the systems' wings would enable free-flight in the atmosphere; fundamentally this is a kite-string set with a wing above and a wing as the resistive anchor set; control of the separate wing set, especially in crosswinding efforts mine the power of winds in different layers of the atmosphere.

Tether Pulling to Drive Generator or Pump Shafts

In the systems of this type, an electrical generator, pump, or tasking line is installed on the ground. There are two subtypes, with or without a secondary vehicle. In the subtype without a secondary vehicle,"Yo-Yo" method, the tether slowly unwinds off a drum on the ground, due to the windward pull of the kite system's wing, while the wing travels crosswind, that is, left-right of the wind's ambient direction, along various paths, e.g., a figure-8 flight path, or optimized lemniscate paths, or circular paths (small or large radius). The turning drum rotates the rotor of the generator or pump through, perhaps, a high-ratio gearbox. Periodically, the wing is depowered, and the tether is reeled in, or, using the crosswind for a constant pull, the tether is re-connected to a different section of the drum while the wing is traveling in a "downwind" cycle. In some systems two tethers are used instead of one.

In another subtype, a secondary vehicle is used. Such a vehicle can be a carousel, a car, railed cart, wheeled land vehicle, or even a ship on the water. The electrical generator is installed on the vehicle. The rotor of the electrical generator is brought in motion by the carousel, the axle of the car, or the screw of the ship, correspondingly.

Onboard Generator

In the systems of this type, one or more flying blades and electrical generators are installed on the wing. The relative airflow rotates the blades by way of autorotation, an interaction with the wind, which transfer the power to the generators. Produced electrical energy is transmitted to the ground through an electrical cable laid along the tether or integrated with the tether. The same blades are sometimes used for double purpose where they are propellers positively driven by costed electricity for launching or special landing or calm-air flight-maintaining purpose.

Fast Motion Transfer with Downwind Ground Receivers

In this type, an electrical generator is installed on the ground and a separate cable or belt, trailing behind the wing, transfers the power to a sprocket on the ground, which rotates the rotor of the generator. The separate belt extends at approximately the speed of the wing. Because of the high speed of that belt, the gearbox is not required.

Motion Transfer with Upwind Ground Receivers

In this type an electrical generator, pump, tasking line set, or lever is installed on the ground upwind of the wing and driven by the operation of two or three or more tethers arranged from a fast-moving crosswinding flying wing set. Examples are found in the research centers of several universities and kite-energy research centers.

Lighter-than-air (LTA) Assisted Twin-coordinating Wing Sets

Several research centers are exploring twin wing sets employing tether pulling of upwind ground-based loads where the crosswinding wing sets use lighter-than-air devices to assure flight in case of lulls in the ambient wind.

LTA-kite-balloon-lofted fast-moving autorating bladed turbine with upwind receiver of electricity

Many in-public-domain patent disclosed teachings and some current research centers are with a focus on using LTA kites to hold bladed turbines using autorotation to drive flown generators.

Flutter-based Crosswind Kite Power Systems, Fast-motion Method

When a wing element in a kite system is designed to have flutter occur, then that fluttering may be harvested for energy to power various loads. In flutter, the wing element travels to crosswind and then reverses to travel to crosswind in a generally opposite direction; the frequency of cycles of reversed direction is high. Flutter in traditional aviation is usually considered a bad and destructive dynamic to be designed out of an aircraft; but in CWKPS, flutter is sometimes designed into the kite system for the specific purpose of converting the wind's kinetic energy to useful purposes; the fast motion of flutter is prized by some kite-energy systems development centers. Harvesting the energy of flutter in kite systems has been done in several ways. One way is to convert the flutter energy into sound, even pleasant sound or music; purposes vary from entertaining one person or a crowd of persons; bird-scaring has been an application. Jerking tether lines by the kite-flown fluttering elements to drive loads to make electricity has been done and is being explored. Pumping fluids by use of flutter-derived energy has been proposed in the kite-energy community. And having the fluttering wing made with appropriate materials and arrangement to be a direct electric-generator part, then electricity can be generated immediately; part of the fluttering wing that is formed to be a magnet flutters by conductive coils forms the parts of the electric generator.

Traction by use of CWKPS

CWKPS are used to move objects immediately over ice, snow, land, ponds, lakes, or oceans. The movement of objects may be done for various reasons: recreation, sport, commerce, industry, science, travel, mine-clearing, defense, offense, plowing, landscaping, etc. The multitude of kite systems flown to crosswind to move kite boarders, land sailors, kite surfers, kite boaters, yachts, ships, catamarans, kayaks, power kiters, kite buggies, kite skiers, kite water skiers, etc., is keeping kite-wing manufacturers busy. SkySails is a leader in saving fuel in the shipping industry by using CWKPS.

Lifters Using CWKPS

In this type of CWKPS the fast-motion of the flying blades or wings harvest the wind's energy to power the lifting capacity of the system. Mass loads ares sometimes close-coupled with the wing set; at other times the mass lifted is distributed along the tether set. A military use of this type involved the rotary-wing kites that appear to be helicopters (but are not) tethered by the kite line; a human observer gets lifted to high points for observation purposes; some of these were used in conjunction with submarine operations with the submarine's towing motion providing the apparent wind for the CWKPS. One example is the Focke-Achgelis Fa 330 Lift-and-place or lift-and-drop uses occur in this type; mass loads are lifted and then placed or dropped; this is done sometimes to overcome barriers or to save ground-transportation fuel costs. When the mass that is lifted is a generator coupled with the crosswinding blades, then the AWES type is changed; this change is the foundation for the focus of some current wind power companies; David Lang is carefully modelling such AWES in coordination with Grant Calverley.

Torque Transfer Over Rotary Tether CWKPS

In this type, rotary cross wind kites drive rings around a guiding tether line. Since the rings are tied together and in tension, torsion can be transferred from the rotating kites to a ground generator. Rotary kite motion around a main lifted tether can rotates either the tether itself, a rotary tether set, or lines fixed across the axis of the main lift tether. On 15 December 2015 this method was the first to successfully complete the someawe. org 100*3 challenge.

Theory

In all types of the crosswind kite power system, the useful power can be approximately described by the Loyd's formula:

$$P = \frac{2}{27} \rho_a A C_L (\frac{C_L}{C_D})^2 V^3$$

where P is power; C_L and C_D are coefficients of lift and drag, respectively; ρ_a is the air density at the altitude of the wing; A is the wing area and V is the wind speed. This formula disregards tether drag, wing and tether weights, change of the air density with altitude and angle of the wing motion vector to the plane, perpendicular to the wind. A more precise formula is:

$$P = \frac{2}{27} \rho_a A C_L G^2 V^3$$

where G is the effective gliding ratio, taking into account the tether drag.

Example: a system with a rigid wing, having dimensions 50 m x 2 m and G=15 in the 12 m/s wind will provide 40 MW of electric power.

Control of Crosswind-kite-power Source

Depending on the final application of a crosswind-kite-power source, appropriate kite control methods are involved. Human control exercised during the full flight session is exampled in crosswind stunt kiting and kiteboarding; the same has been in place for some electricity-producing crosswind-kite-power source, e.g., by Pierre Benhaiem of France. When the crosswind-kite-power source become too large to handle, then companies are building both human-assisted devices and also fully autonomous robotic control systems. Also, there has been demonstrated fully passive crosswind-kite-power source where natural frequencies of a system do permit the absence of human or robot controls; actually, anyone seeing a kited wing toss left and then right in constant motion is seeing a primitive passively controlled crosswind-kite-power source. Advances in computers, sensors, kite steering units, and servo-mechanisms are being applied to attain full autonomy of the launching, flying, and landing of crosswind-kite-power source that are aiming for the utility-scale energy-production market.

Challenges

Some sectors of crosswind kite power are commercially robust already; the sport low altitude traction industry is one of those sectors; toy sport crosswind kite power systems kept at low altitude must remain safe. But the sectors of high altitude larger CWKPS aiming for utility-scale electrical production to compete against other forms of energy production must overcome various challenges to achieve mainstream acceptance. Some of the challenges are regulatory permissions, including use of airspace and land; safety considerations; reliable operation in varying conditions (day, night, summer, winter, fog, high wind, low wind, etc.); third-party assessment and certification; lifecycle cost modeling.

History

The early 1800s witnessed George Pocock using control of kite system wings to crosswind to good effect. In early 1900s Paul Garber would produce high speed wings by two-line controls to give targets for aircraft gunners. Crosswind kite power was brought again into focus when Miles L. Loyd carefully described the mathematics and potential of crosswind kite power in 1980. In 1980 it was not possible to create an economical automatic control system to control the wings of a kite system, though passive control of crosswinding kite systems had been ancient. With the advance of computational and sensory resources fine control of the wings of a kite system became not only affordable, but cheap. In the same time significant progress was made in the materials and wing construction techniques; new types of flexible kites with good L/D ratio have been invented. Synthetic materials suitable for the wing and tether became affordable; among those materials are UHMWPE, carbon fiber, PETE, and rip-stop nylon. A large number of people became engaged in the sports of kitesurfing, kiteboarding, kite buggying, snowkiting, and power kiting. Multiple companies and academic teams work on crosswind kite power. Most of the progress in the field has been achieved in the last 10 years.

Prospects for Crosswind Kite Power

Current trends in CWKPS sectors will have their follow-on stories. Enthusiasm seems to be at a high level among over a thousand workers in the crosswind kite power realm that includes scales from toy scale to utility-grid. Speculation for traveling and moving goods without fuel around the world by use of CWKPS is envisioned both by systems staying connected to the ground and some systems fully disconnected from the ground. Objectives for the future discussed in the literature regard CWKPS facing toy, sport, industry, science, commerce, energy for electrical grid, sailing, and a host of other tasking applications. For CWKPS to compete with solar energy, nuclear energy, fossil fuels, conventional wind power, DWKPS, or other renewable energy sources, the levelized cost of energy from CWKPS will need to become competitive, proven, made known, and adopted; during CWKPS march into the future, other competing sectors will be advancing also. The variety of configurations of kite systems that will fly wings to crosswind for the enhanced power is expected to grow; however, for specific purposes and applications, some winning formats are expected to eventually shine. Placing wing elements that fly to crosswind on huge lofted rope-based arches or even net domes is being researched.

Patents that Involve Crosswind Kite Power

The are two sectors of crosswind kite power patents, those that have placed some technology into the public domain and those that are within protection periods and perhaps have valid claims. Crosswind kite power teachings in each patent are part of what is reviewed by the crosswind kite power research and development community and interested readers.

US 3987987 *Self-erecting windmill* by Charles McCutchen and Peter R. Payne. They filed in January 28, 1975. Their work is now in the public domain.

US8066225 *Multi-tether cross-wind kite power* by Benjamin Tigner filed in Jan 19, 2009, but has a priority date of Jan. 31, 2008. He teaches crosswind kite farming to make electricity.

US6781254 *Windmill kite* by Bryan William Roberts with priority date of Nov 7, 2001. This examples crosswind kite power using flying generators driven by autorotating crosswinding turbine blades which play a second roll of being driven by costing power to fly the aircraft to altitude or bring the aircraft to safe harbor. The fast motion of the crosswinding blades is mined to drive the airborne wind generators at the hub of the rotating blades. This type of machine was featured in Popular Science magazine.

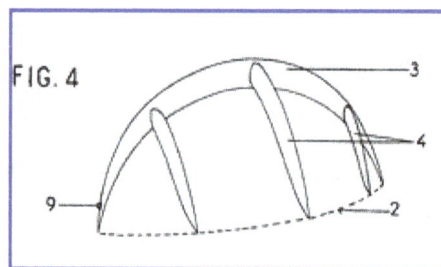

Bruno T. Legaignoux, Dominique M. Legaignoux in their patent show LEI and inflated struts. Wings taught have been being used for crosswind kite power purposes dominantly in kiteboarding and kite surfing, but other uses also/

US4708078 *Propulsive wing with inflatable armature* by Bruno T. Legaignoux, Dominique M. Legaignoux, with priority date of Nov 16, 1984. This patent activity was part of the growing crosswind kite power surge that is still occurring. The inflated leading edge and inflated struts permitted aggressive mining of the wind with crosswind motions and water relauchability. Similar structure technology is being used in some AWES crosswind kite power research centers around the world.

Scale of Crosswind Kite Power Systems

Crosswind kite power systems are found in toy power kites, sport power kites, and experimental-handy sizes; proposed by research centers are huge utility-grid-power-er-feeding sizes. The power gained in toy sizes is used to excite product users; two-line and four-line crosswind toy kite-power systems fill kite festival skies. Serious sport crosswind kite power systems drive the movement of athletes around race courses in local and national competitions. Experimental-handy sizes of crosswind kite power systems are explored while furthering research toward utility-scale systems.

Timeline of uses and Progress of Crosswind Kite Power

Aviation in Britain Before the First World War RAE-0979. This illustrates a non-CWKPS with harvested energy used to lift a man while the wing set refrains from using crosswind dynamics.

Crosswind kite power has been put to various uses throughout history. And the variety of devices that produce crosswind kite power have a historical progression. A simple kite system sitting passively without crosswind kite power production is contrasted with kite systems that fly crosswind producing greater harvesting of energy from the wind's kinetic energy. For perspective, a timeline of crosswind kite power uses and device progress may aid in understanding crosswind kite power.

- 2013: In May 2013, Google acquired a California company developing systems with onboard generators flown to crosswind in circular paths using a hybrid aircraft that double-purposes flying blades as turbine blades as well as costed-energy propellers. Their system is designed to operate as a powered aircraft when needed; the blades and generator are then converted to operate as propellers and motor.

- 2012 November: Progress is exhibited by NTS Gmbh on X-wind (crosswind) kite power system using railed cars that pull line that drives ground generator. NTS X-Wind at EcoSummit Duesseldorf. A closed loop rail with cable-connected cars work in concert to pull the loop cable. Each railed car is pulled by a four-tethered kited wing; each wing is controlled by an autopilot or kite-steering unit.

- 2012 circa: Retail market sees an entry of a crosswind kite power system by Pacific Sky Power. The crosswind elements flown are turbine blades configured in HAWT format with generator aloft at the hub of the turning blades. Their system is not a powered aircraft during any phase of its operation. The scale is of handy one-person mobile size. A pilot lifter kite is used.

- 2010 circa: Making of electricity by using onboard crosswind kite power with the crosswinding elements being autorotating HAWT blades shows by Flygen-Kite under French patent FR 2955627.

- 2009: Airborne Wind Energy Industry Association (AWEIA) formed to serve kite power system industry of all methods including crosswind kite power.

- 1980s: Kiteboarders demonstrate effective upwind travel by use of crosswind kite power techniques.

- Crosswind kite power used in military target practice by Paul E. Garber. The gained crosswind kite power was used to produce speed of the target wing to simulate enemy aircraft.

- 1980: May–June: Miles L. Loyd of Lawrence Livermore National Laboratory, Livermore, California, published in the Journal of Energy, Vol. 4, No. 3, Article: No. 80-4075, Crosswind Kite Power. He focused on flying the wings of kite systems "traverse" to the ambient wind; he noted that the crosswind airspeed of the wings would allow mining the involved kinectic energy for both keeping the wings flying as well as driving other loads for secondary purposes.

- 1820 circa: George Pocock (inventor) demonstrated control of kite-power system to crosswind to obtain energy to draw vehicle rapidly. Many will later consider him as a father of crosswind kite power that uses the harvested wind energy for traction purposes.

Distinguish CWKPS from Non-CWKPS

Kite-power systems dedicated to operating without its energy-harvesting elements flying to crosswind are not CWKPS. Examples help to clarify the two branches of kite-power systems. A simple symmetrical two-stick diamond kite let out to downwind flight while the system's tether pulls to turn an at-ground generator shaft is producing energy for use by flying downwind without flying to crosswind; such is a non-CWKPS. Some hefty downwind kite-power systems (DWKPS) are proposed by serious researchers; some DWKPS instruction is found in the patent literature; one trend involves the opening and closing of pilot-kite-lifted opening-and-closing parachutes to drive generators. Notice that some CWKPS, such as Jalbert parafoil working in figure-8 patterns to turn a ground-stationed generator, could be commissioned to operate fully without flying to crosswind, and the resultant kite-power system would then be a DWKPS. Differently, the CWKPS proposed by users of the autorotating blades stay necessarily as CWKPS. Magenn Power's flip-wing kite-balloon is a DWKPS. Similar flip-wing rotating wings are DWKPS, e.g. that taught in Edwards and Evan patent. Benjamin Franklin's legendary pond-crossing by kite power was a simple DWKPS; he was merely dragged downwind by a downwind-flying kite. A non-CWPKS is historically illustrated by a kite-power harvesting system such as was used by Samuel Franklin Cody for man-lifting with the involved wings set in stable downwind flight without flying to crosswind.

Vertical Axis Wind Turbine

The world's tallest vertical-axis wind turbine, in Cap-Chat, Quebec

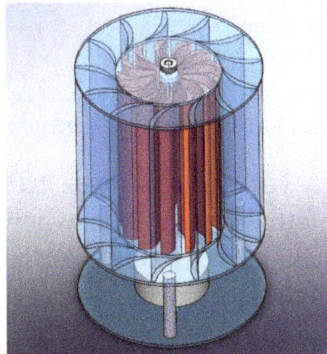

Vortexis Schematic

Vertical-axis wind turbines (VAWTs) are a type of wind turbine where the main rotor shaft is set transverse to the wind (but not necessarily vertically) while the main components are located at the base of the turbine. This arrangement allows the generator and gearbox to be located close to the ground, facilitating service and repair. VAWTs do not need to be pointed into the wind, which removes the need for wind-sensing and orientation mechanisms. Major drawbacks for the early designs (Savonius, Darrieus and giromill) included the significant torque variation or "ripple" during each revolution, and the large bending moments on the blades. Later designs addressed the torque ripple issue by sweeping the blades helically.

A VAWT tipped sideways, with the axis perpendicular to the wind streamlines, functions similarly. A more general term that includes this option is "transverse axis wind turbine" or "cross-flow wind turbine." For example, the original Darrieus patent, US Patent 1835018, includes both options.

Drag-type VAWTs such as the Savonius rotor typically operate at lower tipspeed ratios than lift-based VAWTs such as Darrieus rotors and cycloturbines.

General Aerodynamics

The forces and the velocities acting in a Darrieus turbine are depicted in figure 1. The resultant velocity vector, $\vec{W}$, is the vectorial sum of the undisturbed upstream air velocity, $\vec{U}$, and the velocity vector of the advancing blade, $-\vec{\omega} \times \vec{R}$..

$$\vec{W} = \vec{U} + \left(-\vec{\omega} \times \vec{R}\right)$$

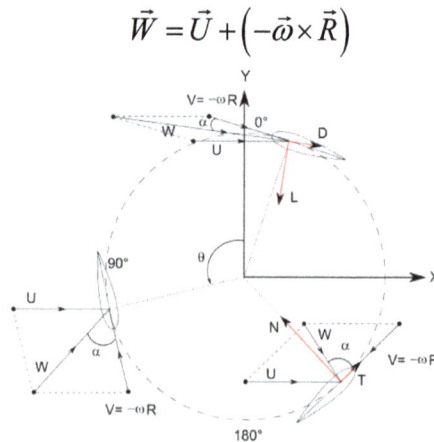

Fig1: Forces and velocities acting in a Darrieus turbine for various azimuthal positions

Thus the oncoming fluid velocity varies during each cycle. Maximum velocity is found for $\theta = 0°$ and the minimum is found for, $\theta = 180°$ where θ is the azimuthal or orbital blade position. The angle of attack, α, is the angle between the oncoming air speed, W, and the blade's chord. The resultant airflow creates a varying, positive angle of attack to the blade in the upstream zone of the machine, switching sign in the downstream zone of the machine.

A helical Darrieus turbine

From geometrical considerations, the resultant airspeed flow and the angle of attack are calculated as follows:

$$W = U\sqrt{1 + 2\lambda\cos\theta + \lambda^2}$$

$$\alpha = \tan^{-1}\left(\frac{\sin\theta}{\cos\theta + \lambda}\right)$$

where $\lambda = \dfrac{\omega R}{U}$ is the tip speed ratio parameter.

The resultant aerodynamic force is resolved either into lift (F_L) - drag (D) components or normal (N) - tangential (T) components. The forces are considered acting at the quarter-chord point, and the pitching moment is determined to resolve the aerodynamic forces. The aeronautical terms "lift" and "drag" refer to the forces across (lift) and along (drag) the approaching net relative airflow. The tangential force acts along the blade's velocity, pulling the blade around, and the normal force acts radially, pushing against the shaft bearings. The lift and the drag force are useful when dealing with the aerodynamic forces around the blade such as dynamic stall, boundary layer etc.; while when dealing with global performance, fatigue loads, etc., it is more convenient to have a normal-tangential frame. The lift and the drag coefficients are usually normalised by the dynamic pressure of the relative airflow, while the normal and tangential coefficients are usually normalised by the dynamic pressure of undisturbed upstream fluid velocity.

$$C_L = \frac{F_L}{1/2\,\rho A W^2}\ ;\ C_D = \frac{D}{1/2\,\rho A W^2}\ ;\ C_T = \frac{T}{1/2\,\rho A U^2 R}\ ;\ C_N = \frac{N}{1/2\,\rho A U^2}$$

A = Blade Area, R = Radius of turbine

The amount of power, P, that can be absorbed by a wind turbine:

$$P = \frac{1}{2}C_p \rho A v^3$$

Where C_p is the power coefficient, ρ is air density, A is the swept area of the turbine, and v is the wind speed.

Advantages of Vertical Axis Wind Turbines

VAWTs offer a number of advantages over traditional horizontal-axis wind turbines (HAWTs).

They are omni-directional and do not need to track the wind. This makes them more reliable due to their not requiring a complex mechanism and motors to yaw the rotor and pitch the blades. In addition, any claimed inefficiencies are in fact cancelled out by the VAWT's ability to take advantage of turbulent and gusty winds. Such winds are not harvested by HAWTs, and in fact cause accelerated fatigue for HAWTs.

The gearbox of a VAWT takes much less fatigue than that of a HAWT. Should it be required, replacement is less costly and simpler, as the gearbox is easily accessible at ground level. This means that a crane or other large equipment is not needed at the site, reducing cost and impact on the environment. Motor and gearbox failures generally increase the operational and maintenance costs of HAWT wind farms both on and offshore.

VAWTs (4Navitas) can use a screw pile foundation, allowing a huge reduction in the carbon cost of an installation as well as a reduction in road transport of concrete during installation. They can be fully recycled at the end of their life.

VAWT wings of the Darrieus type have a constant chord and so are easier to manufacture than the blades of a HAWT, which have a much more complex shape and structure.

VAWTs can be grouped more closely in wind farms, increasing the generated power per unit of land area.

VAWTs can be installed on a wind farm below the existing HAWTs; this will improve the efficiency (power output) of the existing farm.

Research at Caltech has also shown that a carefully designed wind farm using VAWTs can have an output power ten times that of a HAWT wind farm of the same size.

Disadvantages of Vertical Axis Wind Turbines

One of the major outstanding challenges facing vertical axis wind turbine technology is dynamic stall of the blades as the angle of attack varies rapidly.

The blades of a VAWT are fatigue-prone due to the wide variation in applied forces during each rotation. This can be overcome by the use of modern composite materials and improvements in design - including the use of aerodynamic wing tips that cause the spreader wing connections to have a static load. The vertically oriented blades can twist and bend during each turn, causing them to break apart.

VAWTs have proven less reliable than HAWTs. Modern designs of VAWTs have overcome many of the issues associated with early designs.

Applications

The Windspire, a small VAWT intended for individual (home or office) use was developed in the early 2000s by US company Mariah Power. The company reported that several units had been installed across the US by June 2008.

Arborwind, an Ann-Arbor (Michigan, US) based company, produces a patented small VAWT which has been installed at several US locations as of 2013.

In 2011, Sandia National Laboratories wind-energy researchers began a five-year study of applying VAWT design technology to offshore wind farms. The researchers stated: "The economics of offshore windpower are different from land-based turbines, due to installation and operational challenges. VAWTs offer three big advantages that could reduce the cost of wind energy: a lower turbine center of gravity; reduced machine complexity; and better scalability to very large sizes. A lower center of gravity means improved stability afloat and lower gravitational fatigue loads. Additionally, the drivetrain on a VAWT is at or near the surface, potentially making maintenance easier and less time-consuming. Fewer parts, lower fatigue loads and simpler maintenance all lead to reduced maintenance costs."

A 24-unit VAWT demonstration plot was installed in southern California in the early 2010s by Caltech aeronautical professor John Dabiri. His design was incorporated in a 10-unit generating farm installed in 2013 in the Alaskan village of Igiugig.

Dulas, Anglesey received permission in March 2014 to install a prototype VAWT on the breakwater at Port Talbot waterside. The turbine is a new design, supplied by Wales-based C-FEC (Swansea), and will be operated for a two-year trial. This VAWT incorporates a wind shield which blocks the wind from the advancing blades, and thus requires a wind-direction sensor and a positioning mechanism, as opposed to the "egg-beater" types of VAWTs discussed above.

4 Navitas (Blackpool) have been operating two prototype VAWTs since June 2013, powered by a Siemens Power Train, they are due to enter the market in January 2015, with a free technology share to interested parties. 4 Navitas are now in the process of scaling their prototype to 1 MW, (working with PERA Technology) and then floating the turbine on an offshore pontoon. This will reduce the cost of offshore wind energy.

The Dynasphere, is Michael Reynolds' (from Earthship fame) 4th generation vertical axis windmill. These windmills have two 1.5 KW generators and can produce electricity at very low speeds.

Darrieus Wind Turbine

A Darrieus wind turbine once used to generate electricity on the Magdalen Islands

The Darrieus wind turbine is a type of vertical axis wind turbine (VAWT) used to generate electricity from the energy carried in the wind. The turbine consists of a number of curved aerofoil blades mounted on a vertical rotating shaft or framework. The curvature of the blades allows the blade to be stressed only in tension at high rotating speeds. There are several closely related wind turbines that use straight blades. This design of wind turbine was patented by Georges Jean Marie Darrieus, a French aeronautical engineer in 1931. There are major difficulties in protecting the Darrieus turbine from extreme wind conditions and in making it self-starting.

Method of Operation

A very large Darrieus wind turbine on the Gaspé peninsula, Quebec, Canada

Combined Darrieus-Savonius generator used in Taiwan

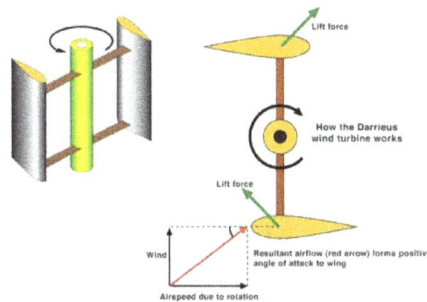
How the Darrieus wind turbine works

In the original versions of the Darrieus design, the aerofoils are arranged so that they are symmetrical and have zero rigging angle, that is, the angle that the aerofoils are set relative to the structure on which they are mounted. This arrangement is equally effective no matter which direction the wind is blowing—in contrast to the conventional type, which must be rotated to face into the wind.

When the Darrieus rotor is spinning, the aerofoils are moving forward through the air in a circular path. Relative to the blade, this oncoming airflow is added vectorially to the wind, so that the resultant airflow creates a varying small positive angle of attack (AoA) to the blade. This generates a net force pointing obliquely forwards along a certain 'line-of-action'. This force can be projected inwards past the turbine axis at a certain distance, giving a positive torque to the shaft, thus helping it to rotate in the direction it is already travelling in. The aerodynamic principles which rotate the rotor are equivalent to that in autogiros, and normal helicopters in autorotation.

As the aerofoil moves around the back of the apparatus, the angle of attack changes to the opposite sign, but the generated force is still obliquely in the direction of rotation, because the wings are symmetrical and the rigging angle is zero. The rotor spins at a rate unrelated to the windspeed, and usually many times faster. The energy arising from the torque and speed may be extracted and converted into useful power by using an electrical generator.

The aeronautical terms lift and drag are, strictly speaking, forces across and along the approaching net relative airflow respectively, so they are not useful here. We really want to know the tangential force pulling the blade around, and the radial force acting against the bearings.

When the rotor is stationary, no net rotational force arises, even if the wind speed rises quite high—the rotor must already be spinning to generate torque. Thus the design is not normally self-starting. Under rare conditions, Darrieus rotors can self-start, so some form of brake is required to hold it when stopped.

One problem with the design is that the angle of attack changes as the turbine spins, so each blade generates its maximum torque at two points on its cycle (front and back of the turbine). This leads to a sinusoidal (pulsing) power cycle that complicates design. In particular, almost all Darrieus turbines have resonant modes where, at a particular rotational speed, the pulsing is at a natural frequency of the blades that can cause them to (eventually) break. For this reason, most Darrieus turbines have mechanical brakes or other speed control devices to keep the turbine from spinning at these speeds for any lengthy period of time.

Another problem arises because the majority of the mass of the rotating mechanism is at the periphery rather than at the hub, as it is with a propeller. This leads to very high centrifugal stresses on the mechanism, which must be stronger and heavier than otherwise to withstand them. One common approach to minimise this is to curve the wings into an "egg-beater" shape (this is called a "troposkein" shape, derived from the Greek for "the shape of a spun rope") such that they are self-supporting and do not require such heavy supports and mountings.

In this configuration, the Darrieus design is theoretically less expensive than a conventional type, as most of the stress is in the blades which torque against the generator located at the bottom of the turbine. The only forces that need to be balanced out vertically are the compression load due to the blades flexing outward (thus attempting to "squeeze" the tower), and the wind force trying to blow the whole turbine over, half of which is transmitted to the bottom and the other half of which can easily be offset with guy wires.

By contrast, a conventional design has all of the force of the wind attempting to push the tower over at the top, where the main bearing is located. Additionally, one cannot easily use guy wires to offset this load, because the propeller spins both above and below the top of the tower. Thus the conventional design requires a strong tower that grows dramatically with the size of the propeller. Modern designs can compensate most tower loads of that variable speed and variable pitch.

In overall comparison, while there are some advantages in Darrieus design there are many more disadvantages, especially with bigger machines in the MW class. The Darrieus design uses much more expensive material in blades while most of the blade is too close to the ground to give any real power. Traditional designs assume that wing tip is at least 40m from ground at lowest point to maximize energy production and lifetime. So far there is no known material (not even carbon fiber) which can meet cyclic load requirements.

Giromills

Fig 3: A Giromill-type wind turbine

Darrieus's 1927 patent also covered practically any possible arrangement using vertical airfoils. One of the more common types is the H-rotor, also called the Giromill or H-bar design, in which the long "egg beater" blades of the common Darrieus design are replaced with straight vertical blade sections attached to the central tower with horizontal supports.

Cycloturbines

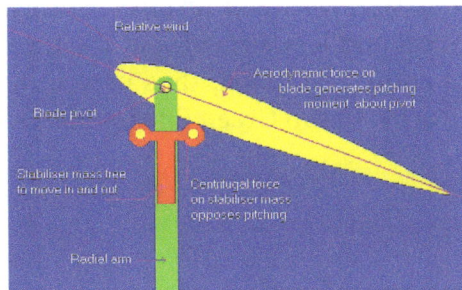

Fig 4: Schematic of mass-stabilised pitch control system.

Another variation of the Giromill is the Cycloturbine, in which each blade is mounted so that it can rotate around its own vertical axis. This allows the blades to be "pitched" so that they always have some angle of attack relative to the wind. The main advantage to this design is that the torque generated remains almost constant over a fairly wide angle, so a Cycloturbine with three or four blades has a fairly constant torque. Over this range of angles, the torque itself is near the maximum possible, meaning that the system also generates more power. The Cycloturbine also has the advantage of being able to self-start, by pitching the "downwind moving" blade flat to the wind to generate drag and start the turbine spinning at a low speed. On the downside, the blade pitching mechanism is complex and generally heavy, and some sort of wind-direction sensor needs to be added in order to pitch the blades properly.

A schematic of a self-acting pitch control system that does not require a wind-direction system is shown in Figure 4.

Helical Blades

A helical Darrieus turbine at Hartnell College.

The blades of a Darrieus turbine can be canted into a helix, e.g. three blades and a helical twist of 60 degrees, similar to Gorlov's water turbines. Since the wind pulls each blade around on both the windward and leeward sides of the turbine, this feature spreads the torque evenly over the entire revolution, thus preventing destructive pulsations. This design is used by the Turby, Urban Green Energy, Enessere and Quiet Revolution brands of wind turbine.

Savonius Wind Turbine

Savonius wind turbines are a type of vertical-axis wind turbine (VAWT), used for converting the force of the wind into torque on a rotating shaft. The turbine consists of a number of aerofoils, usually—but not always—vertically mounted on a rotating shaft or framework, either ground stationed or tethered in airborne systems.

Savonius wind turbine

Origin

The Savonius wind turbine was invented by the Finnish engineer Sigurd Johannes Savonius in 1922. However, Europeans had been experimenting with curved blades on vertical wind turbines for many decades before this. The earliest mention is by the Italian Bishop of Czanad, Fausto Veranzio, who was also an engineer. He wrote in his 1616 book *Machinae novae* about several vertical axis wind turbines with curved or V-shaped blades. None of his or any other earlier examples reached the state of development made by Savonius. In his Finnish biography there is mention of his intention to develop a turbine-type similar to the Flettner-type, but autorotationary. He experimented with his rotor on small rowing vessels on lakes in his country. No results of his particular investigations are known, but the Magnus-Effect is confirmed by König. The two Savonius patents: US1697574, filed in 1925 by Sigurd Johannes Savonius, and US1766765, filed in 1928.

Operation

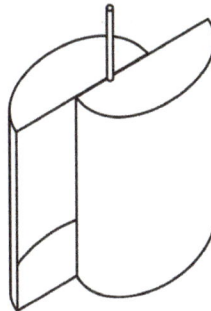

Schematic drawing of a two-scoop Savonius turbine

The Savonius turbine is one of the simplest turbines. Aerodynamically, it is a drag-type device, consisting of two or three scoops. Looking down on the rotor from above, a two-scoop machine would look like an "S" shape in cross section. Because of the curvature, the scoops experience less drag when moving against the wind than when moving with the wind. The differential drag causes the Savonius turbine to spin. Because they are drag-type devices, Savonius turbines extract much less of the wind's power than other similarly-sized lift-type turbines. Much of the swept area of a Savonius rotor may be near the ground, if it has a small mount without an extended post, making the overall energy extraction less effective due to the lower wind speeds found at lower heights.

Power and Rotational Speed

The maximum power of a Savonius rotor is given by $P_{max} = 0.36 \mathrm{kgm}^{-3} \cdot h \cdot r \cdot v^3$, where h and r are the height and radius of the rotor and v is the wind speed.

The angular frequency of a rotor is given by $\omega = \dfrac{\lambda \cdot v}{r}$,, where λ is a dimensionless factor called the tip-speed ratio. The range λ varies within is characteristic of a specific windmill, and for a Savonius rotor λ is typically around ≈ 1.

For example, an oil-barrel sized Savonius rotor with $h=1$ m and $r=0.5$ m under a wind of $v=10$ m/s, will generate a maximum power of 180 W and an angular speed of 20 rad/s (190 revolutions per minute).

Use

Combined Darrieus-Savonius generator in Taiwan

Savonius turbines are used whenever cost or reliability is much more important than efficiency.

Most anemometers are Savonius turbines for this reason, as efficiency is irrelevant to the application of measuring wind speed. Much larger Savonius turbines have been used to generate electric power on deep-water buoys, which need small amounts of power and get very little maintenance. Design is simplified because, unlike with horizontal axis wind turbines (HAWTs), no pointing mechanism is required to allow for shifting wind direction and the turbine is self-starting. Savonius and other vertical-axis machines are good at pumping water and other high torque, low rpm applications and are not usually connected to electric power grids. They can sometimes have long helical scoops, to give smooth torque.

The most ubiquitous application of the Savonius wind turbine is the Flettner Ventilator, which is commonly seen on the roofs of vans and buses and is used as a cooling device. The ventilator was developed by the German aircraft engineer Anton Flettner in the 1920s. It uses the Savonius wind turbine to drive an extractor fan. The vents are still manufactured in the UK by Flettner Ventilator Limited.

Small Savonius wind turbines are sometimes seen used as advertising signs where the rotation helps to draw attention to the item advertised. They sometimes feature a simple two-frame animation.

Tethered Airborne Savonius Turbines

- Airborne wind turbines

- Kite types

- When the Savonius rotor axis is set horizontally and tethered, then kiting results. There are scores of patents and products that use the net lift Magnus-effect that occurs in the autorotation of the Savonius rotor. The spin may be mined for some of its energy for making noise, heat, or electricity.

Gallery

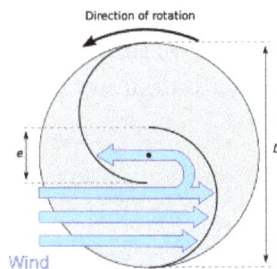

Operation of a Savonius turbine

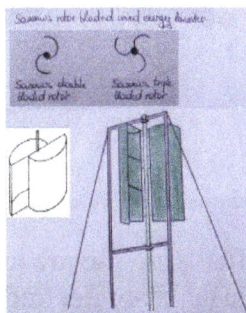

A Savonius rotor bladed WECS

Small Wind Turbine

A small-scale wind tower in rural Indiana, United States

A group of small wind turbines in a community in Dali, Yunnan, China

A small wind turbine is a wind turbine used for microgeneration, as opposed to large commercial wind turbines, such as those found in wind farms, with greater individual power output. The Canadian Wind Energy Association (CanWEA) defines "small wind" as ranging from less than 1000 Watt (1 kW) turbines up to 300 kW turbines. The smaller turbines may be as small as a 50 Watt auxiliary power generator for a boat, caravan, or miniature refrigeration unit.

Design

Evance R9000 5kW small domestic wind turbine with 5.5m rotor diameter

Smaller scale turbines for residential scale use are available. Their blades are usually 1.5 to 3.5 metres (4 ft 11 in–11 ft 6 in) in diameter and produce 1-10 kW of electricity at their optimal wind speed. Some units have been designed to be very lightweight in their construction, e.g. 16 kilograms (35 lb), allowing sensitivity to minor wind movements and a rapid response to wind gusts typically found in urban settings and easy mounting much like a television antenna. It is claimed, and a few are certified, as being inaudible even a few feet (about a metre) under the turbine.

The majority of small wind turbines are traditional horizontal axis wind turbines, but vertical axis wind turbines are a growing type of wind turbine in the small-wind market. Makers of vertical axis wind turbines such as WePower, Urban Green Energy, Helix Wind, and Windspire Energy, have reported increasing sales over the previous years.

The generators for small wind turbines usually are three-phase alternating current generators and the trend is to use the induction type. They are options for direct current output for battery charging and power inverters to convert the power back to AC but at constant frequency for grid connectivity. Some models utilize single-phase generators.

Some small wind turbines can be designed to work at low wind speeds, but in general small wind turbines require a minimum wind speed of 4 metres per second (13 ft/s).

Dynamic braking regulates the speed by dumping excess energy, so that the turbine continues to produce electricity even in high winds. The dynamic braking resistor may be installed inside the building to provide heat (during high winds when more heat is lost by the building, while more heat is also produced by the braking resistor). The location makes low voltage (around 12 volt) distribution practical.

Small units often have direct drive generators, direct current output, lifetime bearings and use a vane to point into the wind. Larger, more costly turbines generally have geared power trains, alternating current output and are actively pointed into the wind. Direct drive generators are also used on some large wind turbines.

Installation

Turbines are often mounted on a tower to raise them above any nearby obstacles. One rule of thumb is that turbines should be at least 30 ft (9.1 m) higher than anything within 500 ft (150 m). Better locations for wind turbines are far away from large upwind obstacles. Measurements made in a boundary layer wind tunnel have indicated that significant detrimental effects associated with nearby obstacles can extend up to 80 times the obstacle's height downwind. However, this is an extreme case. Another approach to siting a small turbine is to use a shelter model to predict how nearby obstacles will affect local wind conditions. Models of this type are general and can be applied to any site. They are often developed based on actual wind measurements, and can estimate flow properties such as mean wind speed and turbulence levels at a potential turbine location, taking into account the size, shape, and distance to any nearby obstacles.

A small wind turbine can be installed on a roof. Installation issues then include the strength of the roof, vibration, and the turbulence caused by the roof ledge. Small-scale rooftop turbines suffer from turbulence and rarely generate significant amounts of power, especially in towns and cities.

Markets

Japan

In July 2012, a new feed-in tariff approved by Japanese Industry Minister Yukio Edano went into effect, promising to boost the country's production of wind and solar energy production. The country is aiming to increase renewable energy investment in part as a response to the Fukushima radiation crisis in March 2011. The feed-in tariff applies to solar panels and small wind turbines and requires utilities to buy back electricity generated from renewable energy sources at government-established rates.

Small-scale wind power (turbines of less than 20 kW capacity) will be subsidized at least 57.75 JPY (about 0.74 USD per kwh).

United Kingdom

Properties in rural or suburban parts of the UK can opt for a wind turbine with inverter to supplement local grid power. The UK's Microgeneration Certification Scheme (MCS) provides feed-in tariffs to owners of qualified small wind turbines.

United States

Small wind turbines added a total of 17.3 MW of generating capacity throughout the United States in 2008, according to the American Wind Energy Association (AWEA). That growth equaled a 78% increase in the domestic market for small wind turbines, which are defined as wind turbines with capacities of 100 kW or less. AWEA's "2009 Small Wind Global Market Study", published in late 2009 May, credited the increase in part to greater manufacturing volumes, as the industry was able to attract enough private investment to finance manufacturing plant expansions. It also credited rising electricity prices and greater public awareness of wind technologies for an increase in residential sale. But a poll of small wind manufacturers found that the growth in 2008 might be only a glimmer of things to come, as the companies projected a 30-fold growth in the U.S. small wind market within as little as five years, despite the global recession. The U.S. small wind industry also benefits from the global market, as it controls about half of the global market share. U.S. manufacturers garnered $77 million of the $156 million that was spent throughout the world on small wind turbine installations. A total of 38.7 MW of small wind power capacity was installed globally in 2008.

In the United States, residential wind turbines with outputs of 2–10 kW, typically cost between US$12,000 and US$55,000 installed (US$6 per watt), although there are incentives and rebates available in 19 states that can reduce the purchase price for home-owners by up to 50 percent, to $3 per watt. The US manufacturer Southwest Windpower estimates a turbine to pay for itself in energy savings in 5 to 12 years.

The dominant models on the market, especially in the United States, are horizontal-axis wind turbines.

To enable consumers to make an informed decision when purchasing a small wind turbine, a method for consumer labeling has been developed by IEA Wind Task 27 in collaboration with IEC TC88 MT2. In 2011 IEA Wind published a Recommended Practice, which describes the tests and procedures required to apply the label.

Croatia

Hybrid system, 2400W windturbines, 4000W solar modules, island Žirje, Croatia

Croatia is an ideal market for small wind turbines due to Mediterranean climate and numerous islands with no access to the electric grid. In winter months when there is less sun, but more wind, small wind turbines are a great addition to isolated renewable energy sites (GSM, stations, marinas etc.). That way solar and wind power provide consistent energy throughout the year.

Germany

In Germany the feed-in tariff for small wind turbines has always been the same as for large turbines. This is the main reason, why the small wind turbine sector in Germany developed slowly. In contrast, small photovoltaic systems in Germany benefited from a high feed-in tariff, at times above 50 Euro-Cent per kilowatt hour.

In August 2014 the German renewable energy law was adjusted, also affecting the feed-in tariffs for wind turbines. For the operation of a small wind turbine with a capacity below 50 kilowatt the tariff amounts to 8.5 Euro-Cent for a period of 20 years.

Due to the low feed-in tariff and high electricity prices in Germany, the economic operation of a small wind turbine depends on a large self-consumption rate of the electricity produced by the small wind turbine. Private households pay on average 28 cent per kilowatt hour for electricity (19% VAT included).

As part of the German renewable energy law 2014 a fee on self-consumed electricity was introduced in August 2014. The regulation does not apply to small power plants with a capacity below 10 kilowatt. With an amount of 1.87 Euro-Cents the fee is low.

DIY Construction

Some hobbyists have built wind turbines from kits, sourced components, or from scratch. DIY wind turbines are usually smaller (rooftop) turbines of approximately 1 kW or less. These small wind turbines are usually tilt-up or fixed / guyed towers.

Do it yourself or DIY-wind turbine construction has been made popular by magazines such as OtherPower and Home Power.

Organizations as Practical Action have designed DIY wind turbines that can be easily built by communities in developing nations and are supplying concrete documents on how to do so.

Floating Wind Turbine

University of Maine's VolturnUS 1:8 was the first grid-connected offshore wind turbine in the Americas.

The world's first full-scale floating wind turbine, Hywind, being assembled in the Åmøy Fjord near Stavanger, Norway in 2009, before deployment in the North Sea

The world's second full-scale floating wind turbine (and first to be installed without the use of heavy-lift vessels), WindFloat, operating at rated capacity (2 MW) approximately 5 km offshore of Aguçadoura, Portugal

A floating wind turbine is an offshore wind turbine mounted on a floating structure that allows the turbine to generate electricity in water depths where bottom-mounted towers are not feasible. Locating wind farms out at sea can reduce visual pollution while providing better accommodation for fishing and shipping lanes. In addition, the wind is typically more consistent and stronger over the sea, due to the absence of topographic features that disrupt wind flow.

Floating wind parks are wind farms that site several floating wind turbines closely together to take advantage of common infrastructure such as power transmission facilities.

History

The concept for large-scale offshore floating wind turbines was introduced by Professor William E. Heronemus at the University of Massachusetts Amherst in 1972. It was not until the mid 1990s, after the commercial wind industry was well established, that the topic was taken up again by the mainstream research community. As of 2003, existing offshore fixed-bottom wind turbine technology deployments had been limited to water depths of 30 metres. Worldwide deep-water wind resources are extremely abundant in subsea areas with depths up to 600 metres, which are thought to best facilitate transmission of the generated electric power to shore communities. Two-thirds of the North Sea is between 50 and 220 meters deep.

In June 2013, the University of Maine made history with its VolturnUS 1:8, a 65-foot-tall floating turbine prototype that is 1:8th the scale of a 6-megawatt (MW), 450-foot rotor diameter design. VolturnUS 1:8 was the first grid-connected offshore wind turbine deployed in the Americas. The VolturnUS design utilizes a concrete semisubmersible floating hull and a composite materials tower designed to reduce both capital and Operation & Maintenance costs, and to allow local manufacturing throughout the US and the World. The VolturnUS technology is the culmination of collaborative research and development conducted by the University of Maine-led DeepCwind Consortium. U.S. Senators Susan Collins and Angus King announced in June 2016 that Maine's New England Aqua Ventus I floating offshore wind demonstration project, designed by the DeepCwind Consortium, has been selected by the U.S. Department of Energy to participate in the Offshore Wind Advanced Technology Demonstration program.

On June 13, 2013, the University of Maine's VolturnUS 1:8 was energized and began delivering electricity through an undersea cable to the Central Maine Power electricity grid, making VolturnUS 1:8 the first grid-connected offshore wind turbine in the Americas.

New England Aqua Ventus I will now be one of up to three leading projects that are each eligible for up to $39.9 million in additional funding over three years for the construction phase of the demonstration program.

Operational Deep-water Platforms

A tension leg mooring system as used by Blue H: left-hand tower-bearing structure (grey) is free floating, the right-hand structure is pulled by the tensioned cables (red) down towards the seabed anchors (light-grey).

In 2011 three floating wind turbine support structures were installed.

Blue H deployed the first 80-kW floating wind turbine 21.3 kilometres (13.2 mi) off the coast of Italy in December 2007. It was then decommissioned at the end of 2008 after completing a planned test year of gathering operational data.

The first large-capacity, 2.3-megawatt floating wind turbine is Hywind, which became operational in the North Sea near Norway in September 2009, and is still operational as of October 2010.

In October 2011, Principle Power's WindFloat Prototype was installed 4 km offshore of Aguçadoura, Portugal in approximately 45 m of water (previously the Aguçadoura Wave Farm site). The WindFloat was fitted with a Vestas V80 2.0-MW offshore wind turbine and grid connected. The installation was the first offshore wind turbine to be deployed without the use of any offshore heavy lift vessels as the turbine was fully commissioned onshore prior to the unit's being towed offshore. This is the first offshore wind turbine installed in open Atlantic waters, and the first to employ a semi-submersible type floating foundation.

SeaTwirl deployed its first floating grid connected wind turbine off the coast of Sweden in August 2011. It was tested and de-commissioned. This design intends to store energy in a flywheel. Thus, energy could be produced even after the wind has stopped blowing.

Blue H Technologies

Blue H Technologies of the Netherlands operated the first floating wind turbine, a prototype deep-water platform with an 80-kilowatt turbine off the coast of Apulia, southeast Italy, in 2008. Installed 21 km off the coast in waters 113 metres deep in order to gather test data on wind and sea conditions, the small prototype unit was decommissioned at the end of 2008.

The Blue H technology utilized a tension-leg platform design and a two-bladed turbine. The two-bladed design can have a "much larger chord, which allows a higher tip speed than those of three-bladers.

As of 2009, Blue H was building a full-scale commercial 2.4-MW$_e$ unit in Brindisi, Italy which it expected to deploy at the same site of the prototype in the southern Adriatic Sea in 2010. This is the first unit in the planned 90-MW *Tricase* offshore wind farm, located more than 20 km off the Puglia coast line.

Hywind

The world's first operational deep-water floating *large-capacity* wind turbine is the Hywind, in the North Sea off Norway. The Hywind was towed out to sea in early June 2009. The 2.3-megawatt turbine was constructed by Siemens Wind Power and mounted on a floating tower with a 100-metre deep draft. The float tower was constructed by Technip. Statoil says that floating wind turbines are still immature and commercialization is distant.

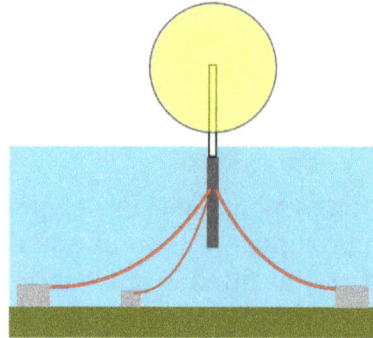

A single floating cylindrical spar buoy moored by catenary cables. Hywind uses a *ballasted catenary* layout that adds 60 tonne weights hanging from the midpoint of each anchor cable to provide additional tension.

The installation is owned by Statoil and will be tested for two years. After assembly in the calmer waters of Åmøy Fjord near Stavanger, Norway, the 120-meter-tall tower with a 2.3-MW turbine was towed 10 km offshore into 220-metre-deep water, 10 km southwest of Karmøy, on 6 June 2009 for a two-year test deployment." Alexandra Beck Gjorv of Statoil said, "[The experiment] should help move offshore wind farms out of sight … The global market for such turbines is potentially enormous, depending on how low we can press costs." The unit became operational in the summer of 2009. Hywind was inaugurated on 8 September 2009. As of October 2010, after a full year of operation, the Hywind turbine is still operating and generating electricity for the Norwegian grid, and still is as of December 2014.

The turbine cost 400 million kroner (around US$62 million) to build and deploy. The 13-kilometre (8.1 mi) long submarine power transmission cable was installed in July, 2009 and system test including rotor blades and initial power transmission was conducted shortly thereafter. The installation is expected to generate about 9 GW·h of electricity annually. The SWATH (Small Waterplane Area Twin Hull) offshore wind turbine service boat, will be tested at Hywind.

Hywind delivered 7.3 GWh in 2010, and survived 11 meter waves with seemingly no wear.

Statoil considers moving the Hywind from Karmøy to a gas platform, reducing gas turbine use.

Hywind 2

As of June 2011, additional pilot Hywind installations were planned in the US and in the North Sea off the coast of Scotland.

In 2013, Statoil pulled out of the $120 million project of four 3-MW turbines floating in 460 feet of water near Boothbay Harbor, Maine citing change in legislation, and focused on their five 6-MW turbines in Scotland instead, where the average wind speed is 10 m/s and the water depth is 100m. The UMaine Aqua Ventus project continues.

In 2015, Statoil received permission to install 30MW Hywinds 18 miles (29 km) outside Peterhead in Scotland, operational around 2017, and plans to test a 1 MWh lithium-ion battery system (called Batwind) with the Hywinds.

Construction of the NOK 2 billion (£152m) project started in 2016 in Spain, Norway and Scotland. Three suction cup anchors will hold each turbine. Plans are to assemble the elements near Stord in summer 2017 and then drag them to Peterhead.

WindFloat

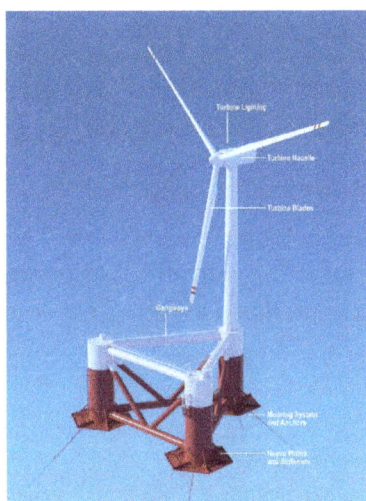

A diagram of the WindFloat system.

WindFloat is a floating foundation for offshore wind turbines designed and patented by Principle Power. A full-scale prototype was constructed in 2011 by Windplus, a joint-venture between EDP, Repsol, Principle Power, A. Silva Matos, Inovcapital, and FAI. The complete system was assembled and commissioned onshore including the turbine. The entire structure was then wet-towed some 400 kilometres (250 mi) (from southern to northern Portugal) to its final installed location 5 kilometres (3.1 mi) offshore of Aguçadoura, Portugal, previously the Aguçadoura Wave Farm. The WindFloat was equipped with a Vestas v80 2.0-megawatt turbine and installation was completed on 22 October 2011. A year later, the turbine had produced 3 GWh.

The subsea metal structure is reported to improve dynamic stability, whilst still maintaining shallow draft, by dampening wave– and turbine–induced motion utilizing a tri-column triangular platform with the wind turbine positioned on one of the three columns. The triangular platform is then "moored" using a conventional catenary mooring consisting of four lines, two of which are connected to the column supporting the turbine, thus creating an "asymmetric mooring."

As the wind shifts direction and changes the loads on the turbine and foundation, a secondary hull-trim system shifts ballast water between each of the three columns. This permits the platform to maintain even keel while producing the maximum amount of energy. This is in contrast to other floating concepts which have implemented control strategies that de-power the turbine to compensate for changes in turbine thrust-induced overturning moment.

This technology could allow wind turbines to be sited in offshore areas that were previously considered inaccessible, areas having water depth exceeding 40 metres and more powerful wind resources than shallow-water offshore wind farms typically encounter.

The cost of this project is around €20 million (about US $26 million). This single wind turbine can produce energy to power 1300 homes.

Principle Power is planning a 30-MW WindFloat project using 6-MW Siemens turbines in 366 metres of water near Coos Bay, Oregon to be operational in 2017.

Sakiyama

In 2010 a 2 MW Hitachi was installed with limited effect near Kabashima. In 2016 it was floated near Fukue Island and run with full effect.

Topologies

Platform topologies can be classified into:

- *single-turbine-floater* (one wind turbine mounted on a floating structure)
- *multiple turbine floaters* (multiple wind turbines mounted on a floating structure)

Engineering Considerations

Undersea mooring of floating wind turbines is accomplished with three principal mooring systems. Two common types of engineered design for anchoring floating structures include tension-leg and catenary loose mooring systems. *Tension leg mooring systems* have vertical tethers under tension providing large restoring moments in pitch and roll. *Catenary mooring systems* provide station–keeping for an offshore structure yet provide little stiffness at low tensions." A third form of mooring system is the *ballasted catenary* configuration, created by adding multiple-tonne weights hanging from the midsection of each anchor cable in order to provide additional cable tension and therefore increase stiffness of the above-water floating structure.

The IEC 61400–3 design standard requires that a loads analysis be based on site-specific external conditions such as wind, wave and currents. The IEC 61400–3-2 standard applies specifically to floating wind turbines.

Economics

The technical feasibility of deepwater floating wind turbines is not questioned, as the long-term survivability of floating structures has been successfully demonstrated by the marine and offshore oil industries over many decades. However, the economics that allowed the deployment of thousands of offshore oil rigs have yet to be demonstrated for floating wind turbine platforms. For deepwater wind turbines, a floating structure will replace pile-driven monopoles or conventional concrete bases that are commonly used as foundations for shallow water and land-based turbines. The floating structure must provide enough buoyancy to support the weight of the turbine and to restrain pitch, roll and heave motions within acceptable limits. The capital costs for the wind turbine itself will not be significantly higher than current marine-proofed turbine costs in shallow water. Therefore, the economics of deepwater wind turbines will be determined primarily by the additional costs of the floating structure and power distribution system, which are offset by higher offshore winds and close proximity to large load centres (e.g. shorter transmission runs).

As of 2009 however, the economic feasibility of shallow-water offshore wind technologies is more completely understood. With empirical data obtained from fixed-bottom installations off many countries since the late 1990s, representative costs are well understood. Shallow-water turbines cost between 2.4 and 3 million United States dollars per megawatt to install, according to the World Energy Council.

As of 2009, the practical feasibility and per-unit economics of deep-water, floating-turbine offshore wind is yet to be established. Initial deployment of single full-capacity turbines in deep-water locations began only in 2009.

As of October 2010, new feasibility studies are supporting that floating turbines are becoming both technically and economically viable in the UK and global energy markets. "The higher up-front costs associated with developing floating wind turbines would be offset by the fact that they would be able to access areas of deep water off the coastline of the UK where winds are stronger and reliable."

The recent Offshore Valuation study conducted in the UK has confirmed that using just one third of the UK's wind, wave and tidal resource could generate energy equivalent to 1 billion barrels of oil per year; the same as North Sea oil and gas production. A significant challenge when using this approach is the coordination needed to develop transmission lines.

In 2014, Statoil decided not to develop Hywind further, but later developed the 30 MW project near Scotland.

A 2015 report by Carbon Trust recommends 11 ways to reduce cost. Also in 2015, researchers at University of Stuttgart estimated cost at €230/MWh.

Oil Well Injection

When oil fields become depleted, the operator injects water to keep pressure high for further extraction. This requires power, but installing gas turbines means shutting down the extraction process, losing valuable income. The classification society DNV GL has calculated that in some cases a floating wind turbine can economically provide power for injection, as the oil platform can keep on producing, avoiding a costly pause.

In 2016 DNV GL, ExxonMobil and others approved calculations of saving $3/barrel of oil using a 6MW Hywind instead of traditional engines, driving two 2MW pumps injecting water into an offshore oil well. At least 44,000 barrels of processed water per day can be injected, even on calm June days.

Floating Design Concepts

Ideol

Ideol's engineers have developed and patented a ring-shaped floating foundation based on a central opening system (Damping Pool) used for optimizing foundation + wind turbine stability. As such, the sloshing water contained in this central opening counteracts the swell-induced floater oscillations. Foundation-fastened mooring lines are simply attached to the seabed to hold the assembly in position. This floating foundation is compatible with all wind turbines without any modification and has reduced dimensions (from 36 to 55 meters per side for a wind turbine between 2 and 8 MW). Manufacturable in concrete or steel, this floating foundation allows for local construction near project sites. Ideol leads the FLOATGEN project, a floating wind turbine demonstration project based on Ideol's technology and planned to be built by Bouygues Travaux Publics and installed by mid-2017 off the coast of Le Croisic on the offshore experimentation site of Ecole Centrale de Nantes (SEM-REV). The construction of this project, France's first offshore wind turbine (precisely 2 MW), is already underway since the 1st of June 2016.

In June 2015, the company has sealed its first commercial contract with the Japanese conglomerate Hitachi Zosen, for the design of the two latest Japanese floating offshore wind demonstrators. In July 2016, Ideol and Hitachi Zosen and Ideol have signed a contract launching the construction phase of their 2 floating offshore wind turbines. These 2 floaters will be each manufactured with different materials (concrete and steel), will be equipped with different wind turbines and will be anchored using different mooring line materials.

In late 2015, Ideol also announced the conclusion of a preliminary collaboration with China Steel Corporation aiming at designing and engineering floating offshore wind turbines.

The French government has recently selected Eolmed, a consortium led by Quadran in association with Ideol, Bouygues Travaux Publics and Senvion, a french renewable energy developer, for the development and construction of a 25MW Mediterranean floating offshore wind farm 15 km off the coastal town of Gruissan (Languedoc-Roussillon).

Nautica Windpower

Nautica Windpower has proposed a technique for potentially reducing system weight, complexity and costs for deepwater sites. Scale model tests in open water have been conducted (September 2007) in Lake Erie and structural dynamics modeling was done in 2010 for larger designs. Nautica Windpower's Advanced Floating Turbine (AFT) uses a single mooring line and a downwind two-bladed rotor configuration that is deflection tolerant and aligns itself with the wind without an active yaw system. Two-bladed, downwind turbine designs that can accommodate flexibility in the blades will potentially prolong blade lifetime, diminish structural system loads and reduce offshore maintenance needs, yielding lower lifecycle costs.

OC3-Hywind

The International Energy Agency (IEA), under the auspices of their *Offshore Code Comparison Collaboration* (OC3) initiative, has completed high-level design and simulation modeling of the *OC-3 Hywind* system, a 5-MW wind turbine installed on a floating spar buoy, moored with catenary mooring lines, in water depth of 320 metres. The spar buoy platform would extend 120 meters below the surface and the mass of such a system, including ballast would exceed 7.4 million kg.

DeepWind

Risø and 11 international partners started a 4-year program called DeepWind in October 2010 to create and test economical floating Vertical Axis Wind Turbines up to 20 MW. The program is supported with €3m through EUs Seventh Framework Programme. Partners include TUDelft, Aalborg University, SINTEF, Statoil and United States National Renewable Energy Laboratory.

VolturnUS

The innovative VolturnUS design utilizes a concrete semisubmersible floating hull and a composite materials tower designed to reduce both capital and Operation & Maintenance costs, and to allow local manufacturing. The VolturnUS technology is the culmination of more than a decade of collaborative research and development conducted by the UMaine Advanced Structures and Composites Center-led DeepCwind Consortium.

North America's first floating grid-connected wind turbine was lowered into the Penobscot River in Maine on 31 May 2013 by the University of Maine Advanced Structures and Composites Center and its partners. The VolturnUS 1:8 was towed down the Penobscot River where it was deployed for 18 months in Castine, ME. During its deployment, it experienced numerous storm events representative of design environmental conditions prescribed by the American Bureau of Shipping (ABS) Guide for Building and Classing Floating Offshore Wind Turbines, 2013.

The patent-pending, VolturnUS floating concrete hull technology can support wind turbines in water depths of 45 meters or more, and has the potential to significantly reduce the cost of offshore wind. With 12 independent cost estimates from around the U.S. and the world, it has been found to significantly reduce costs compared to existing floating systems. The design has also received a complete third-party engineering review.

In June 2016, the UMaine-led New England Aqua Ventus I project won top tier status from the US Department of Energy (DOE) Advanced Technology Demonstration Program for Offshore Wind. This means that the Aqua Ventus project is now automatically eligible for an additional $39.9 Million in construction funding from the DOE, as long as the project continues to meet its milestones. The developer asserts that the Aqua Ventus project will likely become the first commercial scale floating wind project in the Americas.

VertiWind

VertiWind is a floating Vertical Axis Wind Turbine design created by Nenuphar whose mooring system and floater are designed by Technip.

Others

An open source project was proposed by former Siemens director Henrik Stiesdal in 2015 to be assessed by DNV GL. It suggests using tension leg platforms with replaceable pressurized tanks anchored to sheet walls.

Proposals

Floating Wind Farms

As of September 2011, Japan plans to build a pilot floating wind farm, with six 2-megawatt turbines, off the Fukushima coast of northeast Japan where the recent disaster has created a scarcity of electric power. After the evaluation phase is complete in 2016, "Japan plans to build as many as 80 floating wind turbines off Fukushima by 2020." The cost is expected to be in the range of 10–20 billion Yen over five years to build the first six floating wind turbines. Some foreign companies also plan to bid on the 1-GW large floating wind farm that Japan hopes to build by 2020. In March 2012, Japan's Ministry of Economy, Trade and Industry approved a 12.5bn yen ($154m) project to float a 2-MW Fuji in March 2013 and two 7-MW Mitsubishi hydraulic "SeaAngel" later about 20–40 km offshore in 100–150 meters of water depth. The Japanese Wind Power Association claims a potential of 519 GW of floating offshore wind capacity in Japan. The first turbine became operational in November 2013.

As of November 2011, Statoil plans to build a multi-turbine project in Scottish waters utilizing the Hywind design. Hywind was approved for Hawaii in 2016, where two companies also plan to use WindFloats competing for a 400 MW project.

The US State of Maine solicited proposals in September 2010 to build the world's first floating, commercial wind farm. The RFP is seeking proposals for 25 MW of deep-water offshore wind capacity to supply power for 20-year long-term contract period via grid-connected floating wind turbines in the Gulf of Maine. Successful bidders must enter into long-term power supply contracts with either Central Maine Power Company (CMP), Bangor Hydro-Electric Company (BHE), or Maine Public Service Company (MPS). Proposals were due by May 2011.

In April 2012 Statoil received state regulatory approval to build a large four-unit demonstration wind farm off the coast of Maine. As of April 2013, the *Hywind 2* 4-tower, 12–15 MW wind farm was being developed by Statoil North America for placement 20 kilometres (12 mi) off the east coast of Maine in 140–158 metres (459–518 ft)-deep water of the Atlantic Ocean. Like the first Hywind installation off Norway, the turbine foundation will be a *spar floater*. The State of Maine Public Utility Commission voted to approve the construction and fund the US$120 million project by adding approximately 75 cents/month to the average retail electricity consumer. Power could be flowing into the grid no earlier than 2016.

As a result of legislation in 2013 (LD 1472) by the State of Maine, Statoil placed the planned Hywind Maine floating wind turbine development project on hold in July 2013. The legislation required the Maine Public Utilities Commission to undertake a second round of bidding for the offshore wind sites with a different set of ground rules, which subsequently led Statoil to suspend due to increased uncertainty and risk in the project. Statoil is considering other locations for its initial US demonstration project.

Controversy

Some vendors who could bid on the proposed project in Maine expressed concerns in 2010 about dealing with the United States regulatory environment. Since the proposed site is in Federal waters, developers would need a permit from the US Minerals Management Service, "which took more than seven years to approve a yet-to-be-built, shallow-water wind project off Cape Cod", and is also the agency under fire in June 2010 for lax oversight of deepwater oil drilling in Federal waters. "Uncertainty over regulatory hurdles in the United States ... is 'the Achilles heel' for Maine's ambitions for deepwater wind."

Research

Scale modeling and computer modeling attempt to predict the behavior of large–scale wind turbines in order to avoid costly failures and to expand the use of offshore wind power from fixed to floating foundations. Topics for research in this field include:

Computer Models

- Overview of integrated dynamic calculations for floating offshore wind turbines

- Fully coupled aerohydro-servo-elastic response; a basic research tool to validate new designs

Scale Models

- Water tank studies on 1:100 scale Tension-leg Platform and Spar Buoy platforms

- Dynamic response dependency on the mooring configuration

Other Applications

As they are suitable for towing, floating wind turbine units can be relocated to any location on the sea without much additional cost. So they can be used as prototype test units to practically assess the design adequacy and wind power potential of prospective sites.

Floating wind turbines can be used to provide motive power for achieving artificial upwelling of nutrient-rich deep ocean water to the surface for enhancing fisheries growth in areas with tropical and temperate weather. Though deep seawater (below 50 meters depth) is rich in nutrients such as nitrogen and phosphorus, the phytoplankton growth is poor due to the absence of sunlight. The most productive ocean fishing grounds are located in cold water seas at high latitudes where natural upwelling of deep sea water occurs due to inverse thermocline temperatures. The electricity generated by the floating wind turbine would be used to drive high–flow and low–head water pumps to draw cold water from below 50 meters water depth and mixed with warm surface water by eductors before releasing into the sea. Mediterranean sea, Black sea, Caspian sea, Red sea, Persian gulf, deep water lakes/reservoirs are suitable for artificial upwelling for enhancing fish catch economically. These units can also be mobile type to utilise the seasonal favourable winds all around the year.

Unconventional Wind Turbines

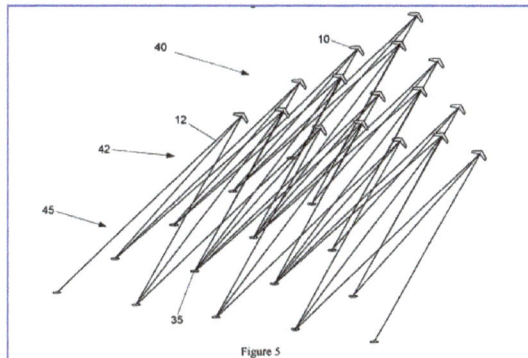

Diagram from United States patent 8066225 illustrates an idea for capturing wind from multiple directions using a large number of tethereed kites.

Unconventional wind turbines are those that differ significantly from the most common types in use. As of 2012, the most common type of wind turbine is the three-bladed upwind horizontal-axis wind turbine (HAWT), where the turbine rotor is at the front of the nacelle and facing the wind upstream of its supporting turbine tower. A second major unit type is also classified by its axis: the vertical-axis wind turbine (VAWT), with blades extending upwards that are supported by a rotating framework.

Due to the large growth of the wind power industry and the length of its historical development dating back to windmills, many different wind turbine designs exist, are in current development, or have been proposed due to their unique features. The wide variety of designs reflects ongoing commercial, technological, and inventive interests in harvesting wind resources both more efficiently and to the greatest extent possible, with costs that may be either lower or greater than conventional three-bladed HAWT designs.

Some turbine designs that differ from the standard type have had limited commercial use, while others have only been demonstrated or are only theoretical concepts with no practical applications. Such unconventional designs cover a wide gamut of innovations, including different rotor types, basic functionalities, supporting structures and form-factors.

Crosswind kite generator with fast motion transfer.

Modified Horizontal

Twin-bladed Rotor

Nearly all modern wind turbines uses rotors with three blades, but some use
only two blades. This was the type used at Kaiser-Wilhelm-Koog, Germany,
where a large experimental two-bladed unit—the GROWIAN, or *Große Wind-
kraftanlage* (big wind turbine)—operated from 1983 to 1987. Other prototypes
and several wind turbine types were also manufactured by NedWind. The Eem-
meerdijk Wind Park in Zeewolde, Netherlands uses only two-bladed turbines.
Wind turbines with two blades are manufactured by Nordic Windpower, such
as model # N 1000, and by GC China Turbine Corp. The NASA wind turbines
(1975-1996) each had 2-blade rotors, producing the same energy at lower cost
than three-blade rotor designs.

Downwind Rotor

Nearly all wind turbines are of an upwind design, meaning the rotor is in front
of the nacelle when the wind is blowing. Some turbines are of a downwind design, meaning the rotor is behind the nacelle when the wind is blowing.

Ducted rotor

Still something of a research project, the ducted rotor consists of a turbine
inside a duct that flares out at the back. They are also referred as Diffus-
er-Augmented Wind Turbines (i.e. DAWT). The main advantage of the ducted rotor is that it can operate in a wide range of winds and generate a higher
power per unit of rotor area. Another advantage is that the generator operates at a high rotation rate, so it doesn't require a bulky gearbox, allowing the mechanical portion to be smaller and lighter. A disadvantage is that
(apart from the gearbox) it is more complicated than the unducted rotor and
the duct is usually quite heavy, which puts an added load on the tower. The
Éolienne Bollée is an example of a DAWT.

Co-axial, multi-rotor

Two or more rotors may be mounted to the same driveshaft, with their combined co-rotation together turning the same generator: fresh wind is brought to each rotor by sufficient spacing between rotors combined with an offset angle (alpha) from the wind direction. Wake vorticity is recovered as the top of a wake hits the bottom of the next rotor. Power has been multiplied several times using co-axial, multiple rotors in testing conducted by inventor and researcher Douglas Selsam, for the California Energy Commission in 2004. The first commercially available co-axial multi-rotor turbine is the patented dual-rotor American Twin Superturbine from Selsam Innovations in California, with 2 propellers separated by 12 feet. It is the most powerful 7-foot-diameter (2.1 m) turbine available, due to this extra rotor. In 2015, Iowa State University aerospace engineers Hui Hu and Anupam Sharma were optimizing designs of multi-rotor systems, including a horizontal-axis co-axial dual-rotor model. In addition to a conventional three-blade rotor, it has a smaller secondary three-blade rotor, covering the near-axis region usually inefficiently harvested. They were considering the overall efficiency of the wind farm, and checking many variations beyond the one mentioned. Preliminary results indicated 10-20% gains, less efficient than is claimed by existing counter-rotating designs but those are complex.

Counter rotating wind turbine (dual rotor)

Counter-rotating horizontal-axis

When a system expels or accelerates mass in one direction, the accelerated mass causes a proportional but opposite force on that system. The spinning blade of a single rotor wind turbine causes a significant amount of tangential or rotational air flow. The energy of this tangential air flow is wasted in a single-rotor propeller design. To use this wasted effort, the placement of a second rotor behind the first takes advantage of the disturbed airflow, and can gain up to 40% more energy from a given swept area as compared with a single rotor. Other advantages of contra-rotation include no gear boxes and auto-centering on the wind (no yaw motors/mechanism required). A patent application dated 1992 exists based on work done with the Trimblemill.

When the counter-rotating turbines are on the same side of the tower, the blades in front are angled forwards slightly so as to avoid hitting the rear ones. If the turbine blades are on opposite sides of the tower, it is best that the blades at the back be smaller than the blades at the front and set to stall at a higher wind speed. This allows the generator to function at a wider wind speed range than a single-turbine generator for a given tower. To reduce sympathetic vibrations, the two turbines should turn at speeds with few common multiples, for example 7:3 speed ratio.

As of 2005, no large practical counter-rotating HAWTs are commercially sold.

Furling tail and twisting blades

> In addition to variable pitch blades, furling tails and twisting blades are other improvements on wind turbines. Similar to the variable pitch blades, they may also greatly increase the efficiency of the turbine and be used in "do-it-yourself" construction

Wind-mill style

> De Nolet is a wind turbine in Rotterdam disguised as a windmill.

Ducted 2-Blade HAWT

> Looking similar to the standard 2-blade or three-blade horizontal-axis wind turbine (HAWT)—the most used types—the British experimented with this type in the early 1950s. As the wind turns the blades, it draws air from near the bottom of the turbine's large hollow mast, and through turbines that spin an electrical generator. Air expels at the tip of the blades. The engineers of this design believed it saved cost by not requiring a linkage and transmission for the generator, and being of lighter weight because the generator was near the bottom of the mast rather than the top. One was built and tested near St Albans, Hertfordshire, England.

Modified Vertical Axis

Aerogenerator

> The Aerogenerator is a special design of vertical axis wind turbine that provided greater energy outputs.

Savonius

The Savonius wind turbine is another special design wind turbine.

Augmented

The "G" Model VAWT Turbine is equipped with three self-positioning Augmentation And Directioning Wings (AADW) placed as the outer sections of classical Darrieus blades. The GMWT can increase almost fivefold the efficiency of classical Darrieus Blades: AADWs adjust themselves to the wind direction without any external power. The resulting combination ("G" Model Wind Turbine) works with very low cut-in wind speed, has self-starting ability, together with a high capacity factor.

Wind Tower Technology

A novel ducted turbine, referred to as a Wind Tower, as a smart architectural integrated design for capturing wind power in either residential or commercial applications is proposed and studied theoretically and experimentally by Navid Goudarzi et al. A Wind Tower uses pressure differentials produced by wind flow around a building to generate electricity. A windcatcher assembly directs the flow into the tower, The tower structure together with the embedded nozzles inside it will accelerate the flow. Different numbers of nozzles together with various inlet and outlet geometries can be studied to obtain an arrangement with an optimum performance at a site. Finally, rotational mechanical power is converted to electrical power using generators. The results show that the Wind Tower technology is a feasible alternative to the practice of using conventional machines in power generation for residential or commercial applications. The advantages of requiring low maintenance, being sustainable and highly reliable make it a great option for off-grid homes and properties and remote areas for power production.

Fuller

The "Fuller" wind turbine is a fully enclosed wind turbine that uses boundary layers instead of blades. Much like a Tesla turbine.

The concept is similar to a stack of disks on a central shaft, separated by a small air gap. The surface tension of air through the small gaps creates friction, making the disks rotate around the shaft. Vanes help direct the air for improved performance, hence it is not totally bladeless.

Aerial

It has been demonstrated that wind turbines could be flown in high-speed winds using high altitude wind power tactics, taking advantage of the winds at high altitudes. A system of automatically controlled tethered kites could also be used to capture energy from high-altitude winds.

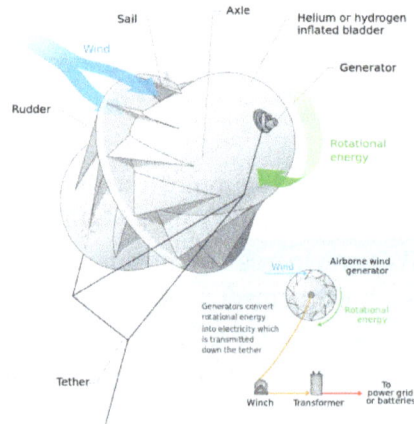

Concept for an airborne wind generator

H-rotor

This is a vertical axis turbine, but it isn't favored because of its poor efficiency. One blade is pushed by the wind while the other is being pushed in the opposite direction. Consequently, only one blade is working at a time.

INVELOX

SheerWind's INVELOX technology was developed by Dr. Daryoush Allaei. The invention is really not a turbine, rather a wind capturing and delivery system to a turbine. In a sense, INVELOX is a wind injection system, much like a fuel injection system for cars. It works by accelerating the wind. A large intake captures wind, funnels it down using tapered pipes leading to a concentrator that ends in a Venturi section and finally wind exits from a diffuser. Turbine(s) are placed inside the Venturi section of the INVELOX. Inside the Venturi the dynamic pressure is very high while the static pressure is low. The Turbine converts dynamic pressure or kinetic energy to mechanical rotation and thereby to electrical power using a generator. The device has been constructed and tested, but also criticized for lack of efficiency.

Saphonian

The Saphonian design produced by Tunisian startup Saphon Energy uses a dish to generate wind pressure and back-and-forth motion that drives a piston.

Windbeam

Zephyr Energy Corporation's patented Windbeam micro generator captures energy from airflow to recharge batteries and power electronic devices. The Windbeam's novel design allows it to operate silently in wind speeds as low as 2 mph. The generator consists of a lightweight beam suspended by durable long-lasting springs within an outer frame. The beam oscillates rapidly when exposed to airflow due to the effects of multiple fluid flow phenomena. A linear alternator assembly converts the oscillating beam motion into usable electrical energy. A lack of bearings and gears eliminates frictional inefficiencies and noise. The generator can operate in low-light environments unsuitable for solar panels (e.g. HVAC ducts) and is inexpensive due to low cost components and simple construction. The scalable technology can be optimized to satisfy the energy requirements and design constraints of a given application.

Wind belt

Invented by Shawn Frayne. A tensioned but flexible belt vibrates by the passing flow of air, due to aeroelastic flutter. A magnet, mounted at one end of the belt translates in and out of coiled windings producing electricity. The company and product are no longer in existence.

Vortex Bladeless

The vortex bladeless device deliberately maximizes vortex shedding, converting wind energy to fluttering of a lightweight vertical pole, then captures that energy with a generator at the bottom of the pole.

Vaneless Ion Wind Generator

Piezoelectric

Another special type of wind turbines are the piezoelectric wind turbines. Turbines with diameters on the scale of 10 centimeters work by flexing piezoelectric crystals as they rotate, sufficient to power small electronic devices.

Traffic-driven

A few proposals call for generating power from the otherwise wasted energy in the draft created by traffic.

Blade Tip Power System (BTPS)

Designed by Imad Mahawili with Honeywell/WindTronics. This design uses many nylon blades and turns a permanent magnet generator inside out. The magnets are on the tips of the blades, and the stator is on the outside of the generator.

Solar Chimney

Wind turbines may also be used in conjunction with a solar collector to extract the energy due to air heated by the Sun and rising through a large vertical Solar updraft tower.

Wind Turbines on Public Display

Kiosk at the base of the Lamma Winds Nordex N50/800kW wind turbine on Lamma Island with displays showing current power output and cumulative energy produced

Most wind turbines around the world belong to individuals or corporations who use them to generate electric power or to perform mechanical work. As such, wind turbines are primarily working devices. However, the large size and height above surroundings of modern industrial wind turbines, combined with their moving rotors, often makes them conspicuous. A few localities have exploited the attention-getting nature of wind turbines, either by putting visitor centers on their bases, or by providing viewing areas. The wind turbines themselves are generally of conventional horizontal-axis, three-bladed design, and generate power to feed electrical grids, but they also serve the unconventional roles of technology demonstration, public relations, and education.

Rooftop Wind-turbines

Wind-turbines can be installed on roofs of buildings, but this is less common than one might expect. Some examples include Marthalen Landi-Silo in Switzerland, Council House 2 in Melbourne, Australia. Ridgeblade in the UK is like a vertical wind turbine on its side mounted on the apex of a pitched roof. While the Ridgeblade is still in the design stage another example like this, already available in France is the Aeolta Aero-Cube. Discovery Tower is an office building in Houston, Texas, that incorporates 10 wind turbines in its architecture.

The Museum of Science in Boston, Massachusetts began constructing a rooftop Wind Turbine Lab in 2009. The lab is testing nine wind turbines from five different manufacturers. Rooftop wind turbines may suffer from turbulence, especially in cities, which reduces power output and accelerates turbine wear. The lab seeks to address the general lack of performance data for urban wind turbines.

Due to structural limitations of buildings, limited space in urban areas, and safety considerations, wind turbines mounted on buildings are usually small (with nameplate capacities in the low kilowatts), rather than the megawatt-class wind turbines that are most economical for wind farms. An exception is the Bahrain World Trade Centre with three 225 kW wind turbines mounted between twin skyscrapers.

References

- Ahmad Y Hassan, Donald Routledge Hill (1986). Islamic Technology: An illustrated history, p. 54. Cambridge University Press. ISBN 0-521-42239-6.

- Alan Wyatt: Electric Power: Challenges and Choices. Book Press Ltd., Toronto 1986, ISBN 0-920650-00-7

- Hau, Erich. "Wind Turbines: Fundamentals, Technologies, Application, Economics" p142. Springer Science & Business Media, 26. feb. 2013. ISBN 3642271510

- Jamieson, Peter. Innovation in Wind Turbine Design sec11-1, John Wiley & Sons, 5 July 2011. Accessed: 26 February 2012. ISBN 1-119-97545-X

- Eric Hau (ed), Wind Turbines Fundamentals, Technologies, Applications, Economics 2nd Edition ,Springer 2006, ISBN 3-540-24240-6 page 121

- Zbigniew Lubosny (2003). Wind Turbine Operation in Electric Power Systems: Advanced Modeling (Power Systems). Berlin: Springer. ISBN 3-540-40340-X.

- Jamieson, Peter. Innovation in Wind Turbine Design p155, John Wiley & Sons, 7 July 2011. Accessed: 26 February 2012. ISBN 0-470-69981-7

- Jamieson, Peter. Innovation in Wind Turbine Design sec9-1, John Wiley & Sons, 7 July 2011. Accessed: 26 February 2012. ISBN 1-119-97612-X

- Hau, E.--(Erich),Snel, Herman (2000). Large wind turbines. Wiley, Chichester, New York. ISBN 0471494569.

- "Iberwind builds on 98% availability with fresh yaw, blade gains". 15 February 2016. Retrieved 30 May 2016.

- Morten Lund (30 May 2016). "Dansk firma sætter prisbelønnet selvhejsende kran i serieproduktion". Ingeniøren. Retrieved 30 May 2016.

- Jeremy Fugleberg (8 May 2014). "Abandoned Dreams of Wind and Light". Atlas Obscura. Retrieved 30 May 2016.

- Tom Gray (11 March 2013). "Fact check: About those 'abandoned' turbines ...". American Wind Energy Association. Retrieved 30 May 2016.

- Small Wind Turbine Purchasing Guide (PDF) (Report). Canadian Wind Energy Association. p. 3-4. Retrieved 1 March 2016.

- Small Wind Turbine Purchasing Guide (PDF) (Report). Canadian Wind Energy Association. p. 6. Retrieved 1 March 2016.

- "DeepCwind Consortium | Advanced Structures & Composites Center | University of Maine". composites.umaine.edu. Retrieved 2016-07-05.

- "Maine Offshore Wind Project Moves to Top Tier of National Offshore Wind Demonstration Program". U.S. Senator Susan Collins. United States Senate. Retrieved 5 July 2016.

- Nilsen, Jannicke. Statoil utvikler eget batteri-system for lagring av vindkraft Teknisk Ukeblad, 21 March 2016. Accessed: 21 March 2016

- Jannicke Nilsen. "Sjekk dimensjonene: Disse kjettingene skal feste Statoils flytende vindmølle til havbunnen". Teknisk Ukeblad. Retrieved 20 April 2016.

- Tormod Haugstad. "Snart kan oljeselskapene få strøm fra flytende havvind - kan spare 3 dollar fatet". Teknisk Ukeblad. Retrieved 4 May 2016.

Wind-Powered Vehicle: An Overview

Wind power can be harnessed using wind turbines to power mechanical vehicles. These vehicles have been dubbed as 'land yachts'. An overview of crafts and vehicles running in the ocean and on land that operate on wind power is provided in this chapter.

Wind-powered Vehicle

Wind-powered vehicles have traditionally been associated with seafaring vehicles that, until the advent of steam engines, relied primarily upon winds which were used to drive the sails of such vehicles to their destinations. In the Western world, such sail-based wind propulsion on water persists in the modern day within primarily leisurely activities, such as sailing boats, sailing ships, yachting, and windsurfing. A special case is ice yachting on ice-covered water.

Terrestrial sail-based wind propulsion in the form of land sailing and land windsurfing are also popular recreational activities.

Terrestrial and seagoing wind propulsion by use of kites as propulsion subassembly are also wind-powered vehicles. OceanKite, KiteShip, KitVes, are just a few contemporary examples of kite-based wind-powered vehicles. Kite buggying is an ongoing wind-powered land vehicle activity.

Competition cars

In 2016, a car from the Technical University of Denmark went faster than its straight headwind.

Wind-powered Mechanical Vehicles

Wind-powered mechanical vehicles primarily use wind turbines installed at a strategic point of the vehicle. The wind power, which is converted into mechanical energy through gears, belts or chains, causes the vehicle to propel forward. While they are not in mainstream use yet, many schools have begun building the new technology and research into their curricula to teach students and to get them active in the subject. Seagoing electric propulsion where the electricity is derived from the kite subassembly is an ongoing activity by KitVes.)

On Land

Terrestrial wind-powered mechanicals includes Ventomobile, and Spirit of Amsterdam (1 & 2). The Mercedes-Benz Formula Zero uses solar cells, batteries, and a sail. The Greenbird, which currently holds a world record for fastest wind-powered vehicle, is sail-powered.

Ventomobile

The InVentus Ventomobile racing at the Aeolus Race 2008

The Ventomobile is a solely wind-powered lightweight three-wheeler designed by University of Stuttgart students. It won the first prize at the Racing Aeolus held at Den Helder, Netherlands, in August 2008. At the Aeolus race, several universities from all over the world participate in race to build the best and fastest wind-powered vehicle.

Matthias Schubert, Chief Technical Officer of the teams' main sponsor REpower Systems AG, applauded the integration of the InVentus Ventomobile project into the coursework of the students: "The achievement of managing a big team over many months, and even making select construction tasks part of undergraduate teaching cannot be estimated highly enough! The enthusiasm the students show in renewable energies and the development of innovative solutions should serve the industry as an example for the development of new technologies."

Spirit of Amsterdam

The wind-powered land vehicles "Spirit of Amsterdam 1" and "Spirit of Amsterdam 2" were built by the Hogeschool van Amsterdam (University of Applied Science Amsterdam). In 2009 & 2010 the Spirit of Amsterdam 1 and 2 won first prize at the Racing Aeolus held in Denmark.

The Spirit of Amsterdam 2 is the second vehicle built by the Hogeschool van, Amsterdam. It uses a wind turbine to capture the wind velocity and uses mechanical power to propel the vehicle against the wind. This vehicle is capable of driving 6.6 meters per second with a 10-meter per second wind. Next to its reduced weight, the main advantage is the onboard computer with its sophisticated control system. This specially designed computer is capable of automatically shifting gear in a fraction of a second, and by this the gears are always shifted to their most efficient position.

Mercedes-Benz Formula Zero

Unlike traditional racing, which focuses merely on the order of finish, Mercedes' new concept introduces energy efficiency as an integral part of the competition. The Formula Zero Racer is loaded with technology designed to extract the maximum thrust from the electric hub motors, aero-efficient solar skin and high-tech rigid sail.

Greenbird

Ecotricity's Greenbird vehicle, designed and piloted by Richard Jenkins, broke the land speed world record for a wind-powered vehicle in 2009."Greenbird recorded a top speed of 126.4 mph (203.4 km/h), and sustained a speed of 126.2 mph (203.1 km/h) for the required time of three seconds, beating the previous, American held, record of 116 mph (186.7 km/h), set by Bob Schumacher in the Iron Duck in March 1999 at the same location.

Blackbird

Land yacht *Blackbird*
Note that the streamers on the yacht and on the ground point in opposite directions

Land yacht *Blackbird*
More recent version, with fairing to improve performance

The Blackbird is an experimental land yacht, built to demonstrate that it is possible to sail directly downwind faster than the wind (DDWFTTW).

In 2006, following a viral internet debate started as a brain teaser, a propeller-driven land yacht was built and filmed, showing it was possible to sail 'dead' downwind faster than the wind by the power of the available wind only.

In 2009, a MIT professor had worked out the equations for such a device and concluded that one could be built in practice "without too much difficulty". Other researchers arrived at similar conclusions.

In the same year, after being challenged that the video was a hoax, team members Rick Cavallaro and John Borton of *Sportvision*, sponsored by Google and in association with the San Jose State University aeronautics department, built a test vehicle nicknamed Blackbird. A year later, in 2010 Cavallaro successfully tested the vehicle, achieving more than 2 times the speed of wind, definitively demonstrating that it is possible to build a vehicle which can achieve the claim.

A second test with an improved vehicle in 2011 reached close to 3 times the speed of wind.

Sailing Ship

ARC Gloria, a three-masted barque and Colombian Navy training ship, at sunset in Cartagena

A modern sailing ship is any large wind-powered vessel. Traditionally a sailing ship (or simply ship) is a sailing vessel that carries three or more masts with square sails on each. Large sailing vessels that are not ship-rigged may be more precisely referred to by their sail rig, such as schooner, barque (also spelled "bark"), brig, barkentine, brigantine or sloop.

Characteristics

USS *Constitution* under sail in Massachusetts Bay, 21 July 1997. Nicknamed "Old Ironsides," the frigate became famous in the War of 1812 when British cannonballs bounced off her oak hull. She is the oldest commissioned warship still afloat, and is open to the public in Charlestown, Massachusetts.

There are many different types of sailing ships, but they all have certain basic things in common. Every sailing ship has a hull, rigging and at least one mast to hold up the sails that use the wind to power the ship.

The crew who sail a ship are called sailors or *hands*. They take turns to take the watch, the active managers of the ship and her performance for a period. Watches are traditionally four hours long. Some sailing ships use traditional ship's bells to tell the time and regulate the watch system, with the bell being rung once for every half hour into the watch and rung eight times at watch end (a four-hour watch).

Ocean journeys by sailing ship can take many months, and a common hazard is becoming becalmed because of lack of wind, or being blown off course by severe storms or winds that do not allow progress in the desired direction. A severe storm could lead to shipwreck, and the loss of all hands.

Sailing ships are limited in their maximum size compared to ships with heat engines, so economies of scale are also limited. The heaviest sailing ships (limited to those vessels for which sails were the primary means of propulsion) never exceeded 14,000 tons displacement. Sailing ships are therefore also very limited in the supply capacity of their holds, so they have to plan long voyages carefully to include many stops to take on provisions and, in the days before watermakers, fresh water.

Types of Sailing Ships

There are many types of sailing ships, mostly distinguished by their rigging, hull, keel, or number and configuration of masts. There are also many types of smaller sailboats not listed here. The following is a list of vessel types, many of which have changed in meaning over time:

- barque, or bark: at least three masts, fore-and-aft rigged mizzen mast

- barquentine: at least three masts with all but the foremost fore-and-aft rigged

- bilander: a ship or brig with a lug-rigged mizzen sail

- brig: two masts square rigged (may have a spanker on the aftermost)

- brigantine: two masts, with the foremast square-rigged

- caravel

- carrack

- catamaran: vessel with two parallel hulls, usually identical or mirror images, linked by beams and deck or "trampoline", with a central mast or hull mounted in rarer circumstances e.g. Team Philips.

- clipper: a square-rigged merchant ship of the 1840–50s designed for speedy passages

- cog: plank built, one mast, square rigged

- corvette: an imprecise term for a small, often ship-rigged vessel

- cutter: Fore-and-aft rigged, single mast with two headsails

- dhow: a lateen-rigged merchant or fishing vessel

- dinghy: a boat with lift-able ballast [centerboard or drop keel], most commonly small and are single masted.

- wangga ndrua or drua, a sacred double hull canoe of Fiji, last made in the 1880s.

- frigate: a ship-rigged European warship with a single gundeck, designed for commerce-raiding and reconnaissance

- fishing smack

- fluyt: a Dutch oceangoing merchant vessel, rigged similarly to a galleon

- full-rigged ship: three or more masts, all of them square rigged

- galleon: a large, primarily square-rigged vessel of the sixteenth and seventeenth centuries

- hermaphrodite brig: similar to a brigantine

- junk: a lug-rigged Chinese tradeship

- ketch: two masts fore-and-aft rigged, the mizzen mast forward of the rudder post

- Koch (boat)

- longship: vessels used by the Vikings, with a single mast and square sail, also propelled by oars.

- lugger: vessel with at least two masts, carrying lugsails

- luzzu

- pink: in the Atlantic, a small oceangoing ship with a narrow stern.

- pram

- schooner: fore-and-aft rigged sails, with two or more masts, the aftermost mast taller or equal to the height of the forward mast(s)

- ship of the line: the largest warship in European navies, ship-rigged

- sloop: a single fore-and-aft rigged mast and bowsprit

- snow: a brig carrying a square mainsail and often a spanker on a trysail mast

- tjotter

- trimaran: vessel with three hulls, the central usually larger, linked by beams and deck.

- wa\a kaulua

- windjammer: large sailing ship with an iron or for the most part steel hull, built to carry cargo in the late nineteenth and early twentieth century.

- xebec: a Mediterranean warship adapted from a galley, with three lateen-rigged masts

- yawl: two masts, fore-and-aft rigged, the mizzen mast aft of the rudder post

- yacht: a small pleasure craft

Automated Sailing

The *Moshulu* is a rare four-masted bark (with square sales on all masts except for the jigger-mast, which carries fore-and-aft sails. She is berthed in Philadelphia and is now open as a restaurant.

In 1902 the sailing vessel *Preussen* was the first to assist handling of sails by making use of steam power without auxiliary engines for propulsion. The steam power was used to drive the winches, hoists and pumps. A similar ship *Kruzenshtern*, a very large sailing vessel without mechanical assists, had a crew of 257 men, compared to the *Preussen*, which required only 48 men.

In 2006, automated control had been taken to the point where sails could be operated by one person using a central control unit, DynaRig. The DynaRig technology was first developed in the 1960s in Germany by W. Prolls as a propulsion alternative for commercial ships to prepare for a possible future energy crisis. The technology is a high-tech version of the same type of sail used by the *Preussen*, the "square-rigger". The main difference is that the yards do not swing around a fixed mast but are rigidly attached to a rotating mast. DynaRig along with extensive computerization was used in the proof-of-concept *Maltese Falcon* to enable it to be sailed with no crew aloft.

As of 2013, with increasing restrictions on use of bunker fuel, attempts were underway to develop hybrid sailing ships using automated sail and alternative fuels.

Rotor Ship

A rotor ship, or Flettner ship, is a type of ship designed to use the Magnus effect for propulsion. The Magnus effect is a force acting on a spinning body in a moving airstream, which acts perpendicularly to the direction of the airstream. In addition to the familiar principle of backspin imparted to increase range in ball sports, the Magnus effect was also employed in the bouncing bombs developed by Barnes Wallis.

Rotor ships typically use rotor sails powered by a motor to take advantage of the effect. German engineer Anton Flettner was the first to build a ship which attempted to tap this force for propulsion.

Rotor ship E-Ship 1, from German wind-turbine manufacturer Enercon.

Background

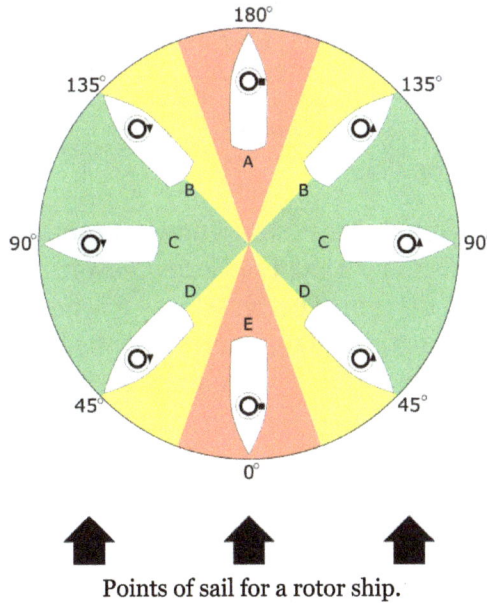

Points of sail for a rotor ship.

A rotor/Flettner ship is a type of ship designed to use the Magnus effect for propulsion. The Magnus effect is a force acting on a spinning body in a moving airstream, which acts perpendicularly to the direction of the airstream. Rotor ships typically use rotor sails — spinning bodies that are essentially vertical cylinders — powered by a motor to take advantage of the effect. These types of propulsion cylinders are now commonly called Flettner rotors. Rotor ships have unique points of sail.

Original Vessels

Invention

The rotor ship *Buckau*.

The rotor ship "Barbara" in Barcelona.

The German engineer Anton Flettner was the first to build a ship which attempted to use the Magnus effect for propulsion.

Assisted by Albert Betz, Jakob Ackeret, and Ludwig Prandtl, Flettner constructed an experimental rotor vessel; October 1924 the Germaniawerft finished construction of a large two-rotor ship named *Buckau*. The vessel was a refitted schooner which carried two cylinders (or rotors) approximately 15 metres (50 ft) high, and 3 metres (10 ft) in diameter, driven by an electric propulsion system of 50 hp (37 kW) power. In 1926, a larger ship with three rotors, the *Barbara,* was built by the shipyard A.G. Weser in Bremen.

Voyages

Following completion of its trials, the *Buckau* set out on her first voyage, from Danzig to Scotland across the North Sea, in February 1925. The rotor ship could tack (sail into the wind) at 20-30 degrees, while a vessel with a typical sail rig, cannot tack closer than 45 degrees to the wind; hence, the rotors did not give cause for concern in stormy weather.

On 31 March 1926, the *Buckau*, now renamed *Baden Baden* after the German spa town, sailed to New York via South America, arriving in New York Harbor on 9 May.

Despite having completed trouble free crossings of the North Sea and Atlantic the power consumed by spinning 15m tall drums was vastly disproportionate to the propulsive effect when compared with conventional screws (propellers). As the Flettner system could not compete economically Flettner turned his attention to other projects, such as his rotor aircraft.

The rotors were removed and the *Baden Baden* returned to screw power, until destroyed in a Caribbean storm in 1931.

Types

Several types of rotor ships can be distinguished. Rotor sail-only ships exist, as do rotor sail-assist (hybrid) ships. Wind Ship Development Corporation has two types of sail-assist designs, for use with different sizes of ships. In practice, most rotor ships have a system with an electric motor that allows the initial start and eventual stop of the rotor by crew. Rotor's rotational speed (i.e., RPM) and direction of spin can also be controlled.

Uses Today

Flensburg catamaran at the Kiel Week 2007.

There has been renewed interest in the rotor ship concept in the later 20th century, starting with Jacques-Yves Cousteau's *Alcyone* in 1983.

The German wind-turbine manufacturer Enercon launched and christened a new rotor ship, on 2 August 2008. The ship is being used to transport turbines and other equipment to locations around the world; the maiden delivery of turbines for Castledockrell Windfarm arrived in Dublin Port on 11th Aug 2010. On 29 July 2013, Enercon provided a press release claiming a potential for "operational fuel savings of up to 25% compared to same-sized conventional freight vessels." after 170,000 sea miles; actual performance figures were not provided.

The University of Flensburg is developing the *Flensburg catamaran* or *Uni-Cat Flensburg*, a rotor-driven catamaran.

The THiiiNK Holding organization describes an improved version of the Flettner Rotor that is retractable and supplemented by an additional control surface. The page claims that "The system has been developed and tested both in tank tests and in full-scale sea trials." The design improvement claims include improved rotor performance (by 50% or more), and an improved internal rate of return (IRR) compared to a standard rotor (up to 55%).

In 2009 the Finland-based maritime engineering company Wärtsilä unveiled a concept for a cruiseferry that would utilise Flettner rotors as means of reducing fuel consumption. This concept has been developed in association with the Finnish ferry operator Viking Line, (the ship, M/S *Viking Grace*, was built 2011–2012, without the rotors). The engine type is also developed by Magnuss since 2011.

In 2007, Stephen H. Salter and John Latham proposed the building of 1,500 robotic rotor ships to mitigate global warming. The ships would spray seawater into the air to enhance cloud reflectivity. A prototype rotor ship was tested on Discovery Project Earth; The rotors were made of carbon fibre and were attached to a retrofitted trimaran and successfully propelled the vessel stably through the water at a speed of six knots. The focus of the experiment was based on the ability for the boat to move emissions free for a specialized purpose, leaving it unclear whether or not the efficiency of the rotors was on parity, inferior to, or superior to conventionally propelled vessels.

E-Ship 1

E-Ship 1 at Cassens Werft shipyard

The *E-Ship 1* is a RoLo cargo ship that made its first voyage with cargo in August 2010. The ship is owned by the third-largest wind turbine manufacturer, Germany's Enercon GmbH. It is used to transport wind turbine components. The *E-Ship 1* is a Flettner ship: four large rotorsails that rise from its deck are rotated via a mechanical linkage to the ship's propellers. The sails, or Flettner rotors, aid the ship's propulsion by means of the Magnus effect – the perpendicular force that is exerted on a spinning body moving through a fluid stream.

Building History

The hull of *E-Ship 1* was built by German shipbuilder Lindenau Werft in Kiel. The ship's launch took place on August 2, 2008, with the delivery date estimated for the first half of 2009. In September 2008, Lindenau-Werft declared insolvency. On January 25, 2009, it was announced that *E-Ship 1* would be towed to and completed by German shipyard Cassens Werft in Emden.

Steel construction work was completed in 2010, and the ship was docked at North Sea Works, where final construction took place with the ship in the water. In April 2010, the *E-Ship 1* returned to Cassens Werft, where preparations were made for sea trials. The ship set off for a first sea trial from Emden to Bremerhaven on 6 July 2010. Three trial runs were completed until the end of July. The ship made its first voyage with cargo in August 2010, carrying nine turbines for Castledockrell Wind Farm from Emden to Dublin, Ireland.

Structure

The ship's bridge is located at the bow, and has three decks and two port-related long-boom cranes with payload capabilities of 80 and 120 tonnes. The ship has a rear ramp, and can function as a RoLo cargo ship. The vessel is 130 meters in length and 22.5 meters wide, with tonnage 12,800 dwt gt/10,000 approx. It is equipped with fore and aft maneuvering thrusters and has an Ice class GL E3 hull rating.

Propulsion and Machinery

E-Ship 1 with Flettner rotors mounted

The *E-Ship 1* is equipped with nine Mitsubishi marine diesel engines with a total output of 3.5 MW. The ship's exhaust gas boilers are connected to a Siemens downstream

steam turbine, which in turn drives four Enercon-developed Flettner rotors. These rotors, resembling four large cylinders mounted on the ship's deck, are 27 meters tall and
4 meters in diameter.

Published Performance Information

Performance results for the ship (including sea-keeping and wind power) was published by Enercon on 23 Sept 2013.

Windmill Ship

A windmill ship, wind energy conversion system ship or wind energy harvester ship
propels itself by use of a windmill to drive a propeller.

They use wind power through a mechanical or electrical transmission to the propeller.
Where transmission is electric, storage batteries may also be used to allow power generated at one time to be used for propulsion later on.

Windmill ships should not be confused with rotor ships, which instead rely on the Magnus effect for propulsion.

Points of Sail

Because a windmill can rotate 360° into the wind, no matter what direction the ship is
facing, a windmill ship can sail in any direction. In fact, because the power produced
depends almost entirely on the apparent wind, they can produce the most power sailing
directly upwind. Note that sailing upwind, while resulting in more power generation
by the wind turbine, requires more power to be expended by the engine and thus it is
still more efficient to sail down wind. To sail upwind, a conventional sailing vessel must
tack across the wind.

Types

Several types can be made; these include windmill-only ships as well as hybrid ships which
store wind power from the windmill when the ship does not need to be propelled. To reduce
the energy required to propel the boat, windmill ships are often equipped with low-friction
hull designs, such as multihulls, or they are hydrofoils. Boats without low-friction hulls or
hydrofoils can be equipped with windmills, but often the force generated by the windmills
alone is not sufficient to propel the craft. In this case, the windmills only provide supplemental force to conventional sails or other propulsion systems.

At present, research is still going on and the best types of bladed rotors still needs to be
determined. For example, high horizontal axis wind mills are proven to make the ship

less stable. Therefore, vertical axis wind mills (e.g. Savonius turbines) are sometimes preferred. Also, the wind mill needs to be highly durable as marine environments tend to degrade windmills more quickly than what is common on land.

Current Ships

Few windmill ships have been built to date; these include:

- Jim Bates' Te whaka
- Lindsay Olen's Thrippence
- Peter Worsley's windmill-driven boat
- Jim Wilkinson's Revelation 2

The film Waterworld starring Kevin Costner featured a trimaran powered by a vertical-axis Darrieus wind turbine.

Wingsail

BMW Oracle Racing USA-17 from the 2010 America's Cup, with a rigid mainsail wingsail, and a conventional jib at the fore

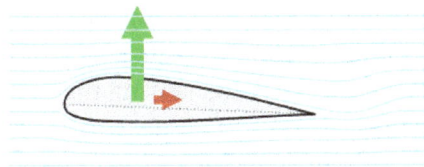

Forces on a wing (green = lift, red = drag) are equivalent on a wingsail.

A wingsail is a variable-camber aerodynamic structure that is fitted to a marine vessel in place of conventional sails. Wingsails are analogous to airplane wings, except that they are designed to provide lift on either side to accommodate being on either tack. Whereas wings adjust camber with flaps, wingsails adjust camber with a flexible or jointed structure (for hard wingsails). Wingsails are typically mounted on an unstayed spar—often made of carbon fiber for lightness and strength. The geometry of wingsails provides more lift, and a better lift-to-drag ratio, than traditional sails. Wingsails are more complex and expensive than conventional sails.

Introduction

Wingsails are of two basic constructions that create an airfoil, "soft" (fabric-shaped) and "hard" (rigid-surfaced). L. Francis Herreshoff pioneered a precursor rig that had jib and main, each with a two-ply sail with leading edges attached to a rotating spar. The C Class Catamaran class has been experimenting and refining wingsails in a racing context since the 60s. Englishman, John Walker, explored the use of wingsails in cargo ships and developed the first practical application for sailing yachts in the 1990s . Wingsails have been applied to small vessels, like the Optimist dinghy and Laser, to cruising yachts, and most notably to high-performance multihull racing sailboats, like USA-17. The smallest craft have a unitary wing that is manually stepped. Cruising rigs have a soft rig that can be lowered, when not in use. High-performance rigs are often assembled of rigid components and must be stepped (installed) and unstepped by shore-side equipment.

Camber Adjustment

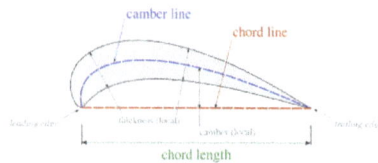

Cross section of an aerofoil showing camber line.

Wingsails change camber (the asymmetry between the top and the bottom surfaces of the aerofoil), depending on tack, and wind speed. A wingsail becomes more efficient with greater curvature towards on the downwind side. Since the windward side changes with each tack, so must sail curvature change. This happens passively on a conventional sail, as it fills in with wind on each tack. On a wingsail, a change in camber requires a mechanism. Wingsails also change camber to adjust for windspeed. On an aircraft flaps increase the camber or curvature of the wing, raising the maximum lift coefficient—the lift a wing can generate—at a lower speeds of air passing over it. A wingsail has the same need for camber adjustment, as windspeed changes—a straighter camber curvature as windspeed increases, more curved as it decreases.

Mechanisms for camber adjustment are similar for soft and hard wingsails. Each employs independent leading and trailing airfoil segments that are adjusted independently for camber. More sophisticated rigs allow for variable adjustment of camber with height above the water to account for increased windspeed.

Comparison with Conventional Sailing Rigs

The presence of rigging, supporting the mast of a conventional fore-and-aft rig limits sail geometry to shapes that are less efficient than the efficient narrow chord of the wingsail. However, conventional sails are simple to adjust for windspeed by reefing. Wingsails typically are a fixed surface area. Conventional sails can be furled easily; some flexible wingsails can be dropped, when not in use; rigid wingsails must be removed when exposure to wind is undesirable.

Points of Sail

Nielsen summarized the efficiencies of wingsails, compared with conventional sails, for different points of sail, as follows:

- Close-hauled: At 30° apparent wind, the wingsail has a 10-degree angle of attack and more lift, compared to the conventional sail plan and its angles of attack of 15° for the jib and 20° for the mainsail.

- Beam reach: At 90° apparent wind, the wingsail, positioned across the boat, functions efficiently as a wing, providing forward lift, whereas the jib of the conventional sail plan suffers from being difficult to shape as a wing (the main sail is still relatively efficient).

- Broad reach: At 135° apparent wind, the wingsail may be eased in such a manner that it still functions efficiently as a wing, whereas the jib and main sail no longer provide lift—instead they present themselves perpendicular to the wind and provide force from drag only.

The top of the wing of an Oracle AC45 racing catamaran

References

- Parker, Dana T. Square Riggers in the United States and Canada, pp. 6-7, Transportation Trails, Polo, IL, 1994. ISBN 0-933449-19-4.

- Parker, Dana T. Square Riggers in the United States and Canada, p. 12, Transportation Trails, Polo, IL, 1994. ISBN 0-933449-19-4.

- Parker, Dana T. Square Riggers in the United States and Canada, p. 25, Transportation Trails, Polo, IL, 1994. ISBN 0-933449-19-4.

- Parker, Dana T. Square Riggers in the United States and Canada, p. 59, Transportation Trails, Polo, IL, 1994. ISBN 0-933449-19-4.

- Parker, Dana T. Square Riggers in the United States and Canada, p. 45, Transportation Trails, Polo, IL, 1994. ISBN 0-933449-19-4.

- Hubert Chanson (30 August 2013). Applied Hydrodynamics: An Introduction. CRC Press. pp. 100–. ISBN 978-1-315-86304-7.

- G. A. Tokaty (20 February 2013). A History and Philosophy of Fluid Mechanics. Courier Corporation. pp. 150–. ISBN 978-0-486-15265-3.

- Albert Einstein (27 September 2011). Essays in Science. Philosophical Library/Open Road. pp. 61–. ISBN 978-1-4532-0479-5.

- Fred M Walker (5 May 2010). Ships and Shipbuilders: Pioneers of Design and Construction. Seaforth Publishing. pp. 220–. ISBN 978-1-84832-072-7.

- G. A. Tokaty (1994). A History and Philosophy of Fluid Mechanics. Courier Corporation. pp. 152–. ISBN 978-0-486-68103-0.

- Ray, Keith. The Strangest Aircraft of All TIme. Stroud, Gloucester GL5 2QG: The History Press. p. 48. ISBN 9780750960977.

- Bahman Zohuri (3 September 2016). Nuclear Energy for Hydrogen Generation through Intermediate Heat Exchangers: A Renewable Source of Energy. Springer. pp. 23–. ISBN 978-3-319-29838-2.

- Houghton, E. L.; Carpenter, P.W. (2003). Butterworth Heinmann, ed. Aerodynamics for Engineering Students (5th ed.). ISBN 0-7506-5111-3.

- Gilmore, C.P. (1984). "Spin Sail: Harnesses Mysterious Magnus Effect for Ship Propulsion," Popular Science (January), pp. 70-73, see [1], accessed 13 October 2015.

- Kennedy, John (2010). "Discovery: State-of-the-art cargo ship to dock with haul of wind turbines". Silicon Republic (online, August, 10). Retrieved 12 October 2015.

- Anon. (2012). "PM E-Ship1 Ergebnisse DBU" (PDF). Enercon.de. Archived from the original (PDF) on June 7, 2014. Retrieved 2015-10-12.

- Nielsen, Peter (May 14, 2014). "Have Wingsails Gone Mainstream?". Sail Magazine. Interlink Media. Retrieved 2015-01-24.

- Widnall, Sheila; Cornwell, Hayden; Williams, Peter (2014), "Effects of Spanwise Flexibility on Lift and Rolling Moment of a Wingsail", Massachusetts Institute of Technology. Department of Aeronautics and Astronautics

- "Cutty Sark," Royal Museums Greenwich Web site (http://www.rmg.co.uk/cuttysark). Retrieved July 29, 2014.

- Latham, John (2007). "Futuristic fleet of 'cloudseeders' (15 February)". BBC. Archived from the original on 2012-07-25. Retrieved 2012-07-25.

- Tony Gray. "Workshop Hints: Ship's Bells". The British Horological Institute. Archived from the original on 9 November 2012. Retrieved 12 June 2011.

- Adam Fischer (February 28, 2011). "One Man's Quest to Outrace Wind". Wired. Retrieved 2012-07-03.

- Cort, Adam (April 5, 2010). "Running Faster than the Wind". sailmagazine.com. Retrieved January 9, 2012.

5

Laws Related to Wind Power

The chapter concentrates on two basic laws related to wind power. Bert's law and wind profile power law; Bert's law directs the maximum power that can be extracted by wind. Wind profile power law is the law that forms a relation between the speed of wind at one height, and the same at another. This section serves as a source to understand the major laws related to wind power.

Betz's Law

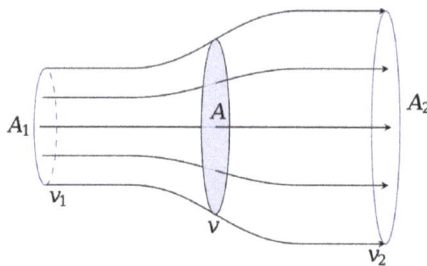

Schematic of fluid flow through a disk-shaped actuator. For a constant density fluid, cross sectional area varies inversely with speed.

Betz's law indicates the maximum power that can be extracted from the wind, independent of the design of a wind turbine in open flow. It was published in 1919, by the German physicist Albert Betz. The law is derived from the principles of conservation of mass and momentum of the air stream flowing through an idealized "actuator disk" that extracts energy from the wind stream. According to Betz's law, no turbine can capture more than 16/27 (59.3%) of the kinetic energy in wind. The factor 16/27 (0.593) is known as Betz's coefficient. Practical utility-scale wind turbines achieve at peak 75% to 80% of the Betz limit.

The Betz limit is based on an open disk actuator. If a diffuser is used to collect additional wind flow and direct it through the turbine, more energy can be extracted, but the limit still applies to the cross-section of the entire structure.

Concepts

Betz' Law applies to all Newtonian fluids, but this article will use wind as an example. Consider that if all of the energy coming from wind movement through a turbine was extracted as useful energy the wind speed afterwards would drop to zero. If the wind stopped moving at the exit of the turbine, then no more fresh wind could get in - it would be blocked. In order to keep the wind moving through the turbine there has to be some wind movement, however small, on the other side with a wind speed greater than zero. Betz' law shows that as air flows through a certain area, and when it slows from losing energy to extraction from a turbine, it must spread out to a wider area. As a result geometry limits any turbine efficiency to 59.3%.

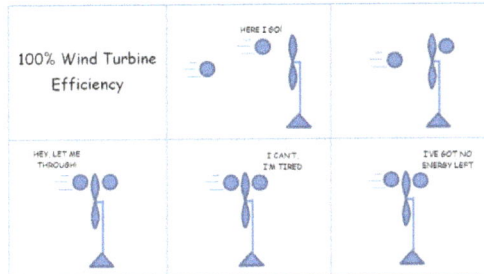

Simple cartoon of two air molecules shows why wind turbines cannot actually run at 100% efficiency.

Three Independent Discoveries of the Turbine Efficiency Limit

The British scientist Frederick W. Lanchester derived the same maximum in 1915. The leader of the Russian aerodynamic school, Nikolay Zhukowsky, also published the same result for an ideal wind turbine in 1920, the same year as Betz did. It is thus an example of Stigler's Law which states that statistically, no scientific discovery is named after its actual discoverer.

Economic Relevance

The Betz limit places an upper bound on the annual energy that can be extracted at a site. Even if a hypothetical wind blew consistently for a full year, no more than the Betz limit of the energy contained in that year's wind could be extracted. In practice, the annual capacity factor of a wind site varies around 25% to 60% of the energy that could be generated with constant wind, limiting the energy that can possibly be obtained even further to typically a range of 14.8% to 35% respectively.

Essentially increasing system economic efficiency results from increased production per unit, measured per square meter of vane exposure. An increase in system efficiency is required to bring down the cost of electrical power production measured in kWh. Efficiency increases may be the result of engineering of the wind capture devices, such as the configuration and dynamics of wind turbines, that may push the power generation from these systems into higher levels of the Betz limit. System efficiency increases in power application, transmission or storage may also contribute to a lower cost of power per unit.

In practicality, most systems do not reach a performance rate of even 50% of the Betz limit, before the further limits of the air stream are ever considered, further lowering the typical rates to 7-17%. Some have claimed to approach the Betz constant and even to surpass it, but none have proven to do so.

Proof

The Betz Limit shows the maximum possible energy that may be derived by means of an infinitely thin rotor from a fluid flowing at a certain speed.

In order to calculate the maximum theoretical efficiency of a thin rotor (of, for example, a windmill) one imagines it to be replaced by a disc that withdraws energy from the fluid passing through it. At a certain distance behind this disc the fluid that has passed through flows with a reduced velocity.

Assumptions

1. The rotor does not possess a hub and is ideal, with an infinite number of blades which have no drag. Any resulting drag would only lower this idealized value.

2. The flow into and out of the rotor is axial. This is a control volume analysis, and to construct a solution the control volume must contain all flow going in and out, failure to account for that flow would violate the conservation equations.

3. The flow is non-compressible. Density remains constant, and there is no heat transfer.

4. Uniform thrust over the disc or rotor area.

Application of Conservation of Mass (Continuity Equation)

Applying conservation of mass to this control volume, the mass flow rate (the mass of fluid flowing per unit time) is given by:

$$\dot{m} = \rho A_1 v_1 = \rho S v = \rho A_2 v_2$$

where v_1 is the speed in the front of the rotor and v_2 is the speed downstream of the rotor, and v is the speed at the fluid power device. ρ is the fluid density, and the area of the turbine is given by S and A_1 and A_2 are the area of the fluid before and after reaching the turbine.

So the density times the area and speed should be equal in each of the three regions, before, while going through the turbine and afterwards.

The force exerted on the wind by the rotor may be written as

$$F = ma$$

$$= m\frac{dv}{dt}$$

$$= \dot{m}\Delta v$$

$$= \rho Sv(v_1 - v_2)$$

or in words, the mass multiplied by the acceleration, so we calculate the air density times the area and speed for the mass and multiply that by the difference in wind speeds before and after for the acceleration.

Power and Work

The work done by the force may be written incrementally as

$$dE = F \cdot dx$$

and the power (rate of work done) of the wind is

$$P = \frac{dE}{dt} = F \cdot \frac{dx}{dt} = F \cdot v$$

Now substituting the force F computed above into the power equation will yield the power extracted from the wind:

$$P = \rho \cdot S \cdot v^2 \cdot (v_1 - v_2)$$

However, power can be computed another way, by using the kinetic energy. Applying the conservation of energy equation to the control volume yields

$$P = \frac{\Delta E}{\Delta t}$$

$$= \frac{1}{2} \cdot \dot{m} \cdot (v_1^2 - v_2^2)$$

Looking back at the continuity equation, a substitution for the mass flow rate yields the following

$$P = \frac{1}{2} \cdot \rho \cdot S \cdot v \cdot (v_1^2 - v_2^2)$$

Both of these expressions for power are completely valid, one was derived by examining the incremental work done and the other by the conservation of energy. Equating these two expressions yields

$$P = \frac{1}{2} \cdot \rho \cdot S \cdot v \cdot (v_1^2 - v_2^2) = \rho \cdot S \cdot v^2 \cdot (v_1 - v_2)$$

Examining the two equated expressions yields an interesting result, namely

$$\frac{1}{2} \cdot (v_1^2 - v_2^2) = \frac{1}{2} \cdot (v_1 - v_2) \cdot (v_1 + v_2) = v \cdot (v_1 - v_2)$$

or

$$v = \frac{1}{2} \cdot (v_1 + v_2)$$

Therefore, the wind velocity at the rotor may be taken as the average of the upstream and downstream velocities. (This is arguably the most counter-intuitive stage of the derivation of Betz' law.)

Betz' Law and Coefficient of Performance

Returning to the previous expression for power based on kinetic energy:

$$\dot{E} = \frac{1}{2} \cdot \dot{m} \cdot \left(v_1^2 - v_2^2 \right)$$

$$= \frac{1}{2} \cdot \rho \cdot S \cdot v \cdot \left(v_1^2 - v_2^2 \right)$$

$$= \frac{1}{4} \cdot \rho \cdot S \cdot (v_1 + v_2) \cdot \left(v_1^2 - v_2^2 \right)$$

$$= \frac{1}{4} \cdot \rho \cdot S \cdot v_1^3 \cdot \left(1 - \left(\frac{v_2}{v_1} \right)^2 + \left(\frac{v_2}{v_1} \right) - \left(\frac{v_2}{v_1} \right)^3 \right).$$

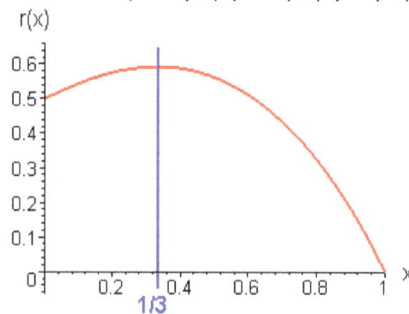

The horizontal axis reflects the ratio v_2/v_1, the vertical axis is the "power coefficient " C_p.

By differentiating $\dot{E}$ with respect to $\dfrac{v_2}{v_1}$ for a given fluid speed v_1 and a given area S one finds the *maximum* or *minimum* value for $\dot{E}$. The result is that $\dot{E}$ reaches maximum value when $\dfrac{v_2}{v_1} = \dfrac{1}{3}$.

Substituting this value results in:

$$P_{\max} = \frac{16}{27} \cdot \frac{1}{2} \cdot \rho \cdot S \cdot v_1^3.$$

The power obtainable from a cylinder of fluid with cross sectional area S and velocity v_1 is:

$$P_{\max} = \frac{16}{27} \cdot \frac{1}{2} \cdot \rho \cdot S \cdot v_1^3.$$

The reference power for the Betz efficiency calculation is the power in a moving fluid in a cylinder with cross sectional area S and velocity v_1:

$$P_{\text{wind}} = \frac{1}{2} \cdot \rho \cdot S \cdot v_1^3.$$

The *"power coefficient"* C_p (= P/P_{wind}) has a maximum value of: $C_{\text{p.max}}$ = 16/27 = 0.593 (or 59.3%; however, coefficients of performance are usually expressed as a decimal, not a percentage).

Modern large wind turbines achieve peak values for C_p in the range of 0.45 to 0.50, about 75% to 85% of the theoretically possible maximum. In high wind speed where the turbine is operating at its rated power the turbine rotates (pitches) its blades to lower C_p to protect itself from damage. The power in the wind increases by a factor of 8 from 12.5 to 25 m/s, so C_p must fall accordingly, getting as low as 0.06 for winds of 25 m/s.

Understanding the Betz Results

Intuitively, the speed ratio of [V_2/V_1 = 0.333] between outgoing and incoming wind, leaving at about a third of the speed it came in, would imply higher losses of kinetic energy. But since a larger area is needed for the now lower density of slower moving (and therefore less pressured) air, energy is conserved.

All energy entering the system is taken into consideration, and local "radial" kinetic energy can have no effect on the outcome, which is the final energy state of the air leaving the system, at a slower speed, larger area and accordingly its lower energy can be calculated.

The last step in calculating the Betz efficiency C_p is to divide the calculated power extracted from the flow by a reference power value. The Betz analysis uses for its power reference, reasonably, the power of air upstream moving at V_1 contained in a cylinder with the cross sectional area of the rotor (S).

Points of Interest

The Betz limit has no dependence on the geometry of the wind extraction system, therefore S may take any form provided that the flow travels from the entrance to the control volume to the exit, and the control volume has uniform entry and exit velocities. Any extraneous effects can only decrease the performance of the system (usually a turbine) since this analysis was idealized to disregard friction. Any non-ideal effects would detract from the energy available in the incoming fluid, lowering the overall efficiency.

Some manufacturers and inventors have made claims of exceeding the Betz' limit by using nozzles and other wind diversion devices, usually by misrepresenting the Betz limit and calculating only the rotor area and not the total input of air contributing to the wind energy extracted from the system.

Modern Development

In 1935 H. Glauert derived the expression for turbine efficiency, when the angular component of velocity is taken into account, by applying an energy balance across the rotor plane. Due to the Glauert model, efficiency is below the Betz limit, and asymptotically approaches this limit when the tip speed ratio goes to infinity.

In 2001, Gorban, Gorlov and Silantyev introduced an exactly solvable model (GGS), that considers non-uniform pressure distribution and curvilinear flow across the turbine plane (issues not included in the Betz approach). They utilized and modified the Kirchhoff model, which describes the turbulent wake behind the actuator as the `degenerated' flow and uses the Euler equation outside the degenerate area. The GGS model predicts that peak efficiency is achieved when the flow through the turbine is approximately 61% of the total flow which is very similar to the Betz result of 2/3 for a flow resulting in peak efficiency, but the GGS predicted that the peak efficiency itself is much smaller: 30.1%.

Recently, viscous computations based on computational fluid dynamics (CFD) were applied to wind turbine modelling and demonstrated satisfactory agreement with experiment. Computed optimal efficiency is, typically, between the Betz limit and the GGS solution.

Wind Profile Power Law

The wind profile power law is a relationship between the wind speeds at one height, and those at another.

The power law is often used in wind power assessments where wind speeds at the height of a turbine (>~ 50 metres) must be estimated from near surface wind observations (~10 metres), or where wind speed data at various heights must be adjusted to a standard height prior to use. Wind profiles are generated and used in a number of atmospheric pollution dispersion models.

The wind profile of the atmospheric boundary layer (surface to around 2000 metres) is generally logarithmic in nature and is best approximated using the log wind profile equation that accounts for surface roughness and atmospheric stability. The wind profile power law relationship is often used as a substitute for the log wind profile when surface roughness or stability information is not available.

The wind profile power law relationship is :

$$\frac{u}{u_r} = \left(\frac{z}{z_r}\right)^\alpha$$

where u is the wind speed (in metres per second) at height z (in metres), and u_r is the known wind speed at a reference height z_r. The exponent (α) is an empirically derived coefficient that varies dependent upon the stability of the atmosphere. For neutral stability conditions, α is approximately 1/7, or 0.143.

In order to estimate the wind speed at a certain height z, the relationship would be rearranged to:

$$u = u_r \left(\frac{z}{z_r}\right)^\alpha$$

The value of 1/7 for α is commonly assumed to be constant in wind resource assessments, because the differences between the two levels are not usually so great as to introduce substantial errors into the estimates (usually < 50 m). However, when a constant exponent is used, it does not account for the roughness of the surface, the displacement of calm winds from the surface due to the presence of obstacles (i.e., zero-plane displacement), or the stability of the atmosphere. In places where trees or structures impede the near-surface wind, the use of a constant 1/7 exponent may yield quite erroneous estimates, and the log wind profile is preferred. Even under neutral stability conditions, an exponent of 0.11 is more appropriate over open water (e.g., for offshore wind farms), than 0.143, which is more applicable over open land surfaces.

Wind Power Density

Estimates of wind power density are presented as wind class, ranging from 1 to 7. The speeds are average wind speeds over the course of a year, although the frequency distribution of wind speed can provide different power densities for the same average wind speed.

Class	10 m (33 ft)		30 m (98 ft)		50 m (164 ft)	
	Wind power density (W/m²)	Speed m/s (mph)	Wind power density (W/m²)	Speed m/s (mph)	Wind power density (W/m²)	Speed m/s (mph)
1	0 - 100	0 - 4.4 (0 - 9.8)	0 - 160	0 - 5.1 (0 - 11.4)	0 - 200	0 - 5.6 (0 - 12.5)
2	100 - 150	4.4 - 5.1 (9.8 - 11.5)	160 - 240	5.1 - 5.9 (11.4 - 13.2)	200 - 300	5.6 - 6.4 (12.5 - 14.3)
3	150 - 200	5.1 - 5.6 (11.5 - 12.5)	240 - 320	5.9 - 6.5 (13.2 - 14.6)	300 - 400	6.4 - 7.0 (14.3 - 15.7)
4	200 - 250	5.6 - 6.0 (12.5 - 13.4)	320 - 400	6.5 - 7.0 (14.6 - 15.7)	400 - 500	7.0 - 7.5 (15.7 - 16.8)
5	250 - 300	6.0 - 6.4 (13.4 - 14.3)	400 - 480	7.0 - 7.4 (15.7 - 16.6)	500 - 600	7.5 - 8.0 (16.8 - 17.9)
6	300 - 400	6.4 - 7.0 (14.3 - 15.7)	480 - 640	7.4 - 8.2 (16.6 - 18.3)	600 - 800	8.0 - 8.8 (17.9 - 19.7)
7	400 - 1000	7.0 - 9.4 (15.7 - 21.1)	640 - 1600	8.2 - 11.0 (18.3 - 24.7)	800 - 2000	8.8 - 11.9 (19.7 - 26.6)

References

- Beychok, Milton R. (2005). Fundamentals Of Stack Gas Dispersion (4th ed.). author-published. ISBN 0-9644588-0-2.

- Manwell, J. F.; McGowan, J. G.; Rogers, A. L. (February 2012). Wind Energy Explained: Theory, Design and Application. Chichester, West Sussex, UK: John Wiley & Sons Ltd. pp. 92–96. ISBN 9780470015001.

- Dodgy wind? Why "innovative" turbines are often anything but Mike Barnard, Gizmag online magazine, June 4, 2013

6

Wind Power: Future Prospects

Wind power can be utilized in several ways in the near coming future. Some of the future prospects discussed in the following chapter are wind power forecasting, wind hybrid power systems and wind resource assessment. Wind power forecast provides an estimation of the expected production of one or more wind turbines. Wind power is emerging as a technology; the following chapter will not provide an overview, it will also delve into the topics related to it.

Wind Power Forecasting

A wind power forecast corresponds to an estimate of the expected production of one or more wind turbines (referred to as a wind farm) in the near future. By production is often meant available power for wind farm considered (with units kW or MW depending on the wind farm nominal capacity). Forecasts can also be expressed in terms of energy, by integrating power production over each time interval.

Time Scales of Forecasts

Forecasting of the wind power generation may be considered at different time scales, depending on the intended application. From milliseconds up to a few minutes, forecasts can be used for the turbine active control. Such type of forecasts are usually referred to as *very short-term* forecasts. For the following 48–72 hours, forecasts are needed for the power system management or energy trading. They may serve for deciding on the use of conventional power plants (Unit commitment) and for the optimization of the scheduling of these plants (Economic dispatch). Bids for energy to be supplied on a day are usually required during the morning of the previous day. These forecasts are called *short-term* forecasts. For longer time scales (up to 5–7 days ahead), forecasts may be considered for planning the maintenance of wind farms, or conventional power plants or transmission lines. Maintenance of offshore wind farms may be particularly costly, so optimal planning of maintenance operations is of particular importance.

For the last two possibilities, the temporal resolution of wind power predictions ranges between 10 minutes and a few hours (depending on the forecast length). Improvements of wind power forecasting has focused on using more data as input to the models involved, and on providing uncertainty estimates along with the traditionally provided predictions.

Reason for Wind Power Forecasts

In the electricity grid at any moment balance must be maintained between electricity consumption and generation - otherwise disturbances in power quality or supply may occur. Wind generation is a direct function of wind speed and, in contrast to conventional generation systems, is not easily dispatchable. Fluctuations of wind generation thus receive a great amount of attention. Variability of wind generation can be regarded at various time scales. First, wind power production is subject to seasonal variations, i.e. it may be higher in winter in Northern Europe due to low-pressure meteorological systems or it may be higher in summer in the Mediterranean regions owing to strong summer breezes. There are also daily cycles which may be substantial, mainly due to daily temperature changes. Finally, fluctuations are observed at the very short-term scale (at the minute or intra-minute scale). The variations are not of the same order for these three different timescales. Managing the variability of wind generation is the key aspect associated to the optimal integration of that renewable energy into electricity grids.

The challenges to face when wind generation is injected in a power system depend on the share of that renewable energy. It is a basic concept, the wind penetration which allows one to describe the share of wind generation in the electricity mix of a given power system. For Denmark, which is a country with one of the highest shares of wind power in the electricity mix, the average wind power penetration over the year is of 16-20% (meaning that 16-20% of the electricity consumption is met wind energy), while the instantaneous penetration (that is, the instantaneous wind power production compared to the consumption to be met at a given time) may be above 100%.

The transmission system operator (TSO) is responsible for managing the electricity balance on the grid: at any time, electricity production has to match consumption. Therefore, the use of production means is scheduled in advance in order to respond to load profiles. The load corresponds to the total electricity consumption over the area of interest. Load profiles are usually given by load forecasts which are of high accuracy. For making up the daily schedule, TSOs may consider their own power production means, if they have any, and/ or they can purchase power generation from Independent Power Producers (IPPs) and utilities, via bilateral contracts or electricity pools. In the context of deregulation, more and more players appear on the market, thus breaking the traditional situation of vertically-integrated utilities with quasi local monopolies. Two main mechanisms compose electricity markets. The first one is the spot market where participants propose quantities of energy for the following day at a given production cost. An auction system permits to settle the electricity spot price for the various periods depending on the different bids. The second mechanism is the balancing of power generation, which is coordinated by the TSO. Depending on the energy lacks and surplus (e.g. due to power plant failures or to intermittence in the case of wind power installations), the TSO determines the penalties that will be paid by IPPs who missed in their obligations. In some cases, an intra-day market is also present, in order to take corrective actions.

In order to illustrate this electricity market mechanism, let us consider the Dutch electricity market. Market participants, referred to as Program Responsible Parties (PRPs), submit their price-quantity bids before 11 am for the delivery period covering the following day from midnight to midnight. The Program Time Unit (PTU) on the balancing market is of 15 minutes. Balancing of the 15-minute averaged power is required from all electrical producers and consumers connected to the grid, who for this purpose may be organised in sub-sets. Since these sub-sets are referred to as Programmes, balancing on the 15-minute scale is referred to as Programme Balance. Programme Balance now is maintained by using the production schedules issued the day before delivery and measurement reports (distributed the day after delivery). When the measured power is not equal to the scheduled power, the *Programme Imbalance* is the difference between the realised sum of production and consumption and the forecast sum of production and consumption. If only production from wind energy is taken into account, Programme Imbalance reduces to realised wind production minus forecast wind production. The programme imbalance is the wind production forecast error.

Programme Imbalance is settled by the System Operator, with different tariffs for negative Programme Imbalance and positive Programme Imbalance. A positive Programme Imbalance indicates more energy actually produced than forecast. by wind energy the realised wind production is bigger than the forecast wind production. And vice versa, in the case of a negative Programme Imbalance by wind energy.

Note that the costs for positive and negative imbalances may be asymmetric, depending on the balancing market mechanism. In general, wind power producers are penalized by such market system since a great part of their production may be subject to penalties.

In parallel to be used for market participation, wind power forecasts may be used for the optimal combined operation of wind and conventional generation, wind and hydro-power generation, or wind in combination with some energy storage devices. They also serve as a basis for quantifying the reserve needs for compensating the eventual lacks of wind production.

General Methodology

Several methods are used for short-term prediction of wind generation. The simplest ones are based on climatology or averages of past production values. They may be considered as reference forecasting methods since they are easy to implement, as well as benchmark when evaluating more advanced approaches. The most popular of these reference methods is certainly *persistence*. This naive predictor — commonly referred to as 'what you see is what you get' — states that the future wind generation will be the same as the last measured value. Despite its apparent simplicity, this naive method might be hard to beat for look-ahead times up to 4–6 hours ahead

Advanced approaches for short-term wind power forecasting necessitate predictions of meteorological variables as input. Then, they differ in the way predictions of meteorological variables are converted to predictions of wind power production, through the so-called *power curve*. Such advanced methods are traditionally divided into two groups. The first group, referred to as physical approach, focuses on the description of the wind flow around and inside the wind farm, and use the manufacturer's power curve, for proposing an estimation of the wind power output. In parallel the second group, referred to as statistical approach, concentrates on capturing the relation between meteorological predictions (and possibly historical measurements) and power output through statistical models whose parameters have to be estimated from data, without making any assumption on the physical phenomena.

Prediction of Meteorological Variables

Wind power generation is directly linked to weather conditions and thus the first aspect of wind power forecasting is the prediction of future values of the necessary weather variables at the level of the wind farm. This is done by using numerical weather prediction (NWP) models. Such models are based on equations governing the motions and forces affecting motion of fluids. From the knowledge of the actual state of the atmosphere, the system of equations allows to estimate what the evolution of state variables, e.g. temperature, velocity, humidity and pressure, will be at a series of grid points. The meteorological variables that are needed as input for wind power prediction obviously include wind speed and direction, but also possibly temperature, pressure and humidity. The distance between grid points is called the spatial resolution of the NWPs. The mesh typically has spacing that varies between few kilometers and up to 50 kilometers for mesoscale models. Regarding the time axis, the forecast length of most of the operational models today is between 48 and 172 hours ahead, which is in adequacy with the requirements for the wind power application. The temporal resolution is usually between 1 and 3 hours. NWP models impose their temporal resolution to short-term wind power forecasting methods since they are used as a direct input.

Predictions of meteorological variables are provided by meteorological institutes. Meteorologists employ atmospheric models for weather forecasts on short and medium term periods. An atmospheric model is a numerical approximation of the physical description of the state of the atmosphere in the near future, and usually is run on a supercomputer. Each computation starts with initial conditions originating from recent measurements. The output consists of the expected instantaneous value of physical quantities at various vertical levels in a horizontal grid and stepping in time up to several hours after initiation. There are several reasons why atmospheric models only approximate reality. First of all, not all relevant atmospheric processes are included in the model. Also, the initial conditions may contain errors (which in a worse case propagate), and the output is only available for discrete points in space (horizontal as well as vertical) and time. Finally, the initial conditions age with time - they are already old when the computation starts let alone when the output is published. Predictions of meteorological variables are issued several times per day (commonly between 2 and 4 times per day), and are available few hours after the beginning of the forecast period. This is because some time is needed for acquiring and analyzing the wealth of measurements used as input to NWP models, then run the model and check and distribute the output forecast series. This gap is a blind spot in the forecasts from an atmospheric model. As an example in the Netherlands, KNMI publishes 4 times per day expected values of wind speed, wind direction, temperature and pressure for the period the between 0 and 48 hours after initialization of the atmospheric model Hirlam with measured data, and then the period before forecast delivery is of 4 hours.

Many different atmospheric models are available, ranging from academic research tools to fully operational instruments. Besides for the very nature of the model (physical processes or numerical schemes) there are some clear distinctive differences between them: time domain (from several hours to 6 days ahead), area (several 10.000 km² to an area covering half the planet), horizontal resolution (1 km to 100 km) and temporal resolution (1 hour to several hours).

One of the atmospheric models is the High Resolution Limited Area Model, abbreviated HiRLAM, which is frequently used in Europe. HiRLAM comes in many versions, that's why it is better to speak about "a" HiRLAM rather than "the" HiRLAM. Each version is maintained by a national institute such as the Dutch KNMI, the Danish DMI or Finnish FMI. And each institute has several versions under her wing, divided into categories such as: operational, pre-operational, semi operational and for research purposes.

Other atmospheric models are UKMO in the UK, Lokalmodell in Germany, Alladin in France (Alladin and Lokalmodell are also used by some other country's within Europe), and MM5 in the USA.

Physical Approach to Wind Power Forecasting

Meteorological forecasts are given at specific nodes of a grid covering an area. Since wind farms are not situated on these nodes, it is then needed to extrapolate these forecasts at the desired location and at turbine hub height. Physical-based forecasting methods consist of several sub-models which altogether deliver the translation from the wind forecast at a certain grid point and model level, to power forecast at the site considered. Every sub-model contains the mathematical description of the physical processes relevant to the translation. Knowledge of all relevant processes is therefore crucial when developing a purely physical prediction method (such as the early versions of the Danish Prediktor). The core idea of physical approaches is to refine the NWPs by using physical considerations about the terrain such as the roughness, orography and obstacles, and by modeling the local wind profile possibly accounting for atmospheric stability. The two main alternatives to do so are: *(i)* to combine the modeling of the wind profile (with a logarithmic assumption in most of the cases) and the geostrophic drag law for obtaining surface winds; *(ii)* to use a CFD (Computational Fluid Dynamics) code that allows one to accurately compute the wind field that the farm will see, considering a full description of the terrain.

When the wind at the level of the wind farm and at hub height is known, the second step consists in converting wind speed to power. Usually, that task is carried out with theoretical power curves provided by the wind turbine manufacturer. However, since several studies have shown the interest of using empirically derived power curve instead of theoretical ones, theoretical power curves are less and less considered. When applying a physical methodology, the modeling of the function which gives the wind generation from NWPs at given locations around the wind farm is done once for all. Then, the estimated transfer function is consequently applied to the available weather predictions at a given moment. In order to account for systematic forecasting errors that may be due to the NWP model or to their modeling approach, physical modelers often integrate Model Output Statistics (MOS) for post-processing power forecasts.

Statistical Approach to Wind Power Forecasting

Statistical prediction methods are based on one or several models that establish the relation between historical values of power, as well as historical and forecast values of meteorological variables, and wind power measurements. The physical phenomena are not decomposed and accounted for, even if expertise of the problem is crucial for choosing the right meteorological variables and designing suitable models. Model parameters are estimated from a set of past available data, and they are regularly updated during online operation by accounting for any newly available information (i.e. meteorological forecasts and power measurements).

Statistical models include linear and non-linear models, but also structural and black-box types of models. Structural models rely on the analyst's expertise on the phenomenon of interest while black-box models require little subject-matter knowledge and are constructed from data in a fairly mechanical way. Concerning wind power forecasting, structural models would be those that include a modeling of the diurnal wind speed variations, or an explicit function of meteorological variable predictions. Black-box models include most of the artificial-intelligence-based models such as Neural-Networks (NNs) and Support Vector Machines (SVMs). However, some models are 'in-between' the two extremes of being completely black-box or structural. This is the case of expert systems, which learn from experience (from a dataset), and for which prior knowledge can be injected. We then talk about grey-box modeling. Statistical models are usually composed by an autoregressive part, for seizing the persistent behavior of the wind, and by a 'meteorological' part, which consists in the nonlinear transformation of meteorological variable forecasts. The autoregressive part permits to significantly enhance forecast accuracy for horizons up to 6–10 hours ahead, i.e. over a period during which the sole use of meteorological forecast information may not be sufficient for outperforming persistence.

Today, major developments of statistical approaches to wind power prediction concentrate on the use of multiple meteorological forecasts (from different meteorological offices) as input and forecast combination, as well as on the optimal use of spatially distributed measurement data for prediction error correction, or alternatively for issuing warnings on potentially large uncertainty.

Uncertainty of Wind Power Forecasts

Current designs are optimal only for nonturbulent, steady conditions. Design tools accounting for unsteadiness and turbulence are far less developed.

Predictions of wind power output are traditionally provided in the form of point forecasts, i.e. a single value for each look-ahead time, which corresponds to the expectation or most-likely outcome. They have the advantage of being easily understandable because this single value is expected to tell everything about future power generation. Today, a major part of the research efforts on wind power forecasting still focuses on point prediction only, with the aim of assimilating more and more observations in the models or refining the resolution of physical models for better representing wind fields at the very local scale for instance. These efforts may lead to a significant decrease of the level of prediction error.

However, even by better understanding and modeling both the meteorological and power conversion processes, there will always be an inherent and irreducible uncertainty in every prediction. This epistemic uncertainty corresponds to the incomplete knowledge one has of the processes that influence future events. Therefore, in complement to point forecasts of wind generation for the coming hours or days, of major importance is to provide means for assessing online the accuracy of these predictions. In practice today, uncertainty is expressed in the form of probabilistic forecasts or with risk indices provided along with the traditional point predictions. It has been shown that some decisions related to wind power management and trading are more optimal when accounting for prediction uncertainty. For the example of the trading application, studies have shown that reliable estimation of prediction uncertainty allows wind power producer to significantly increase their income in comparison to the sole use of an advanced point forecasting method. Other studies of this type deal with optimal dynamic quantification of reserve requirements, optimal operation of combined systems including wind, or multi-area multi-stage regulation. More and more research efforts are expected on prediction uncertainty and related topics.

There are a number of questions that have still yet to be answered, according to a report from a coalition of researchers from universities, industry, and government, supported by the Atkinson Center for a Sustainable Future at Cornell University. They include:

- How do wind farms with their multiple wakes interact with the atmospheric boundary layer to determine the net power that can be produced?

- How do uneven terrain, roughness of the land or sea surface, and turbulence above the boundary layer and turbine wakes affect unsteady loading of downstream wind turbine blades?

- What is the effect of the atmospheric stability (convective, neutral, or stably stratified) on the performance and loading characteristics throughout a typical daily cycle?

- What is the optimal placement of wind turbines in an array, so that the kinetic energy capture can be maximized and unsteady loading be minimized?

The report also provides possible tools used to support this necessary research.

Accuracy

The correlation between wind output and prediction can be relatively high, with an average uncorrected error of 8.8% in Germany over a two-year period.

Wind Hybrid Power Systems

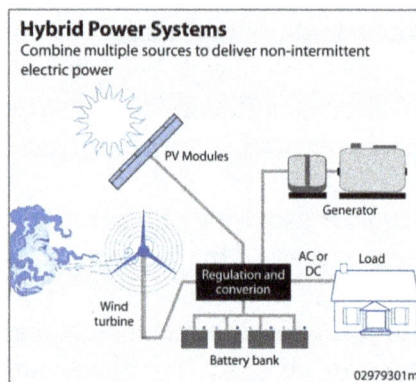

A hybrid wind and solar power system

Wind hybrid power systems combines wind turbines with other storage and/or generation sources. One of the key issues with wind energy is its intermittent nature. This has led to numerous methods of storing energy.

Wind-hydro System

A wind-hydro system generates electric energy combining wind turbines and pumped storage. The combination has been the subject of long-term discussion, and an experimental plant, which also tested wind turbines, was implemented by Nova Scotia Power at its Wreck Cove hydro electric power site in the late 1970s, but was decommissioned within ten years. Since, no other system has been implemented at a single location as of late 2010.

Wind-hydro stations dedicate all, or a significant portion, of their wind power resources to pumping water into pumped storage reservoirs. These reservoirs are an implementation of grid energy storage.

Advantages

Wind and its generation potential is inherently variable. However, when this energy source is used to pump water into reservoirs at an elevation (the principle behind pumped storage), the potential energy of the water is relatively stable and can be used to generate electrical power by releasing it into a hydropower plant when needed. The combination has been described as particularly suited to islands that are not connected to larger grids.

Proposals

During the 1980s, an installation was proposed in the Netherlands. The IJsselmeer would be used as the reservoir, with wind turbines located on its dike. Feasibility studies have been conducted for installations on the island of Ramea (Newfoundland and Labrador) and on the Lower Brule Indian Reservation (South Dakota).

An installation at Ikaria Island, Greece, had entered the construction phase as of 2010.

The island of El Hierro is where the first world's first wind-hydro power station is expected to be complete. Current TV called this "a blueprint for a sustainable future on planet Earth". It is designed to cover between 80-100% of the island's power and is set to be operational in 2012.

Wind-hydrogen System

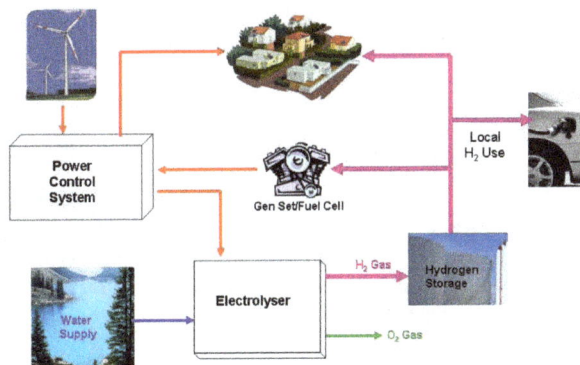

One method of storing wind energy is the production of hydrogen through the electrolysis of water. This hydrogen is subsequently used to generate electricity during periods when demand can not be matched by wind alone. The energy in the stored hydrogen can be converted into electrical power through fuel cell technology or a combustion engine linked to an electrical generator.

Successfully storing hydrogen has many issues which need to be overcome, such as embrittlement of the materials used in the power system.

This technology is being developed in many countries and has even seen a recent IPO of an Australian firm called Wind Hydrogen that looks to commercialise this technology in both Australia and the UK. Essentially Wind Hydrogen offers a source of domestic and vehicular energy for rural communities where current energy transmission costs are prohibitive. Test sites include:

Community	Country	Wind MW
Ramea, Newfoundland and Labrador	Newfoundland, Canada	0.3
Prince Edward Island Wind-Hydrogen Village	PEI, Canada	
Lolland	Denmark	
Bismarck	North Dakota, US	
Koluel Kaike	Santa Cruz, Argentina	
Ladymoor Renewable Energy Project (LREP)	Scotland	
Hunterston Hydrogen Project	Scotland	
RES2H2	Greece	0.50
Unst	Scotland	0.03
Utsira	Norway	0.60

Wind-diesel System

Wind Diesel system on Ramea in Canada

A wind-diesel hybrid power system combines diesel generators and wind turbines, usually alongside ancillary equipment such as energy storage, power converters, and various control components, to generate electricity. They are designed to increase capacity and reduce the cost and environmental impact of electrical generation in remote communities and facilities that are not linked to a power grid. Wind-diesel hybrid systems reduce reliance on diesel fuel, which creates pollution and is costly to transport.

History

Wind-diesel generating systems have been under development and trialled in a number of locations during the latter part of the 20th century. A growing number of viable sites have been developed with increased reliability and minimized technical support costs in remote communities.

Technology

The successful integration of wind energy with diesel generating sets relies on complex controls to ensure correct sharing of intermittent wind energy and controllable diesel generation to meet the demand of the usually variable load. The common measure of performance for wind diesel systems is Wind Penetration which is the ratio between Wind Power and Total Power delivered, e.g. 60% wind penetration implies that 60% of the system power comes from the wind. Wind Penetration figures can be either peak or long term. Sites such as Mawson Station, Antarctica, as well as Coral Bay and Bremer Bay in Australia have peak wind penetrations of around 90%. Technical solutions to the varying wind output include controlling wind output using variable speed wind turbines (e.g. Enercon, Denham, Western Australia), controlling demand such as the heating load (e.g. Mawson), storing energy in a flywheel (e.g. Powercorp, Coral Bay). Some installations are now being converted to wind hydrogen systems such as on Ramea in Canada which is due for completion in 2010.

Communities Using Wind-diesel Hybrids

The following is a, probably incomplete, list of isolated communities utilizing commercial Wind-Diesel hybrid systems with a significant proportion of the energy being derived from wind.

Community	Country	Diesel (in MW)	Wind (in MW)	Population	Date Commissioned	Wind Penetration (peak)	Notes
Mawson Station	Antarctica	0.48	0.60		2003	>90%	
Ross Island	Antarctica	3	1		2009	65%	
Bremer Bay	Australia	1.28	0.60	240	2005	>90%	
Cocos	Australia	1.28	0.08	628			
Coral Bay	Australia	2.24	0.60		2007	93%	
Denham	Australia	2.61	1.02	600	1998	>70%	
Esperance	Australia	14.0	5.85		2003		
Hopetoun	Australia	1.37	0.60	350	2004	>90%	

King Island	Australia	6.00	2.50	2000	2005	100%	Currently (2013) expanding to include 2 MW Diesel-UPS, 3 MW / 1.6 MWh Advanced Lead Acid battery and dynamic load control through smart grid
Rottnest Island	Australia	0.64	0.60		2005		
Thursday Island, Queensland	Australia		0.45	?			
Ramea	Canada	2.78	0.40	600	2003		Being converted to Wind Hydrogen
Sal	Cape Verde	2.82	0.60		2001	14%	
Mindelo	Cape Verde	11.20	0.90			14%	
Alto Baguales	Chile	16.9	2.00	18,703	2002	20%	4.6 MW hydro
Dachen Island	China	1.30	0.15			15%	
San Cristobal, Galapagos Island	Ecuador		2.4		2007		Expanding to cover 100% of island's energy needs by 2015
Berasoli	Eritrea	0.08	0.03				Under tender
Rahaita	Eritrea	0.08	0.03				Under tender
Heleb	Eritrea	0.08	0.03				Under tender
Osmussaar	Estonia	?	0.03		2002		
Kythnos	Greece	2.77	0.31				
Lemnos	Greece	10.40	1.14				

La Désirade	Guade-loupe	0.88	0.14			40%	
Sagar Island	India	0.28	0.50				
Marsabit	Kenya	0.30	0.15			46%	
Frøya	Norway	0.05	0.06			100%	
Batanes	Philip-pines	1.25	0.18		2004		
Flores Island	Portugal		0.60			60%	
Graciosa Island	Portugal	3.56	0.80			60%	
Cape Clear	Ireland	0.07	0.06	100	1987	70%	
Chukotka	Russia	0.5	2.5				
Fuerteven-tura	Spain	0.15	0.23				
Saint Helena	UK		0.48		1999 – 2009	30%	
Foula	UK	0.05	0.06	31		70%	
Rathlin Island	UK	0.26	0.99			100%	
Toksook Bay, Alaska	United States	1.10	0.30	500	2006		
Kasigluk, Alaska	United States	1.10	0.30	500	2006		
Wales, Alaska	United States		0.40	160	2002	100%	
St. Paul, Alaska	United States	0.30	0.68			100%	
Kotzebue, Alaska	United States	11.00			1999	35%	
Savoonga, Alaska	United States		0.20		2008		
Tin City, Alaska	United States		0.23		2008		
Nome, Alaska	United States		0.90		2008		
Hooper Bay, Alaska	United States		0.30		2008		

Wind-diesel Hybrids at Mining Sites

Recently, in Northern Canada wind-diesel hybrid power systems were built by the mining industry. In remote locations in Lac de Gras and Katinniq, Ungava Peninsula, Nunavik two systems are used to save fuel at mines. There is another system in Argentina.

Wind-compressed Air Systems

At power stations that use compressed air energy storage (CAES), electrical energy is used to compress air and store it in underground facilities such as caverns or abandoned mines. During later periods of high electrical demand, the air is released to power turbines, generally using supplemental natural gas. Power stations that make significant use of CAES are operational in McIntosh, Alabama, Germany, and Japan. System disadvantages include some energy losses in the CAES process; also, the need for supplemental use of fossil fuels such as natural gas means that these systems do not completely make use of renewable energy.

The Iowa Stored Energy Park, projected to begin commercial operation in 2015, will use wind farms in Iowa as an energy source in conjunction with CAES.

Wind-solar Systems

Horizontal axis wind-turbine, combined with a solar panel on a lighting pylon at Weihai, Shandong province, China

Wind-solar Building

The Pearl River Tower in Guangzhou, China, will mix solar panel on its windows and several wind turbines at different stories of its structure, allowing this tower to be energy positive.

Wind-solar Lighting

In several parts of China, there are lighting pylons with combinations of solar panels and wind-turbines at their top. This allows space already used for lighting to be used more efficiently with two complementary energy productions units. Most common models use horizontal axis wind-turbines, but now models are appearing with vertical axis wind-turbines, using a helicoidal shaped, twisted-Savonius system.

Wind Resource Assessment

Wind resource assessment is the process by which wind power developers estimate the future energy production of a wind farm. Accurate wind resource assessments are crucial to the successful development of wind farms.

History

Modern wind resource assessments have been conducted since the first wind farms were developed in the late 1970s. The methods used were pioneered by developers and researchers in Denmark, where the modern wind power industry first developed.

Wind Resource Maps

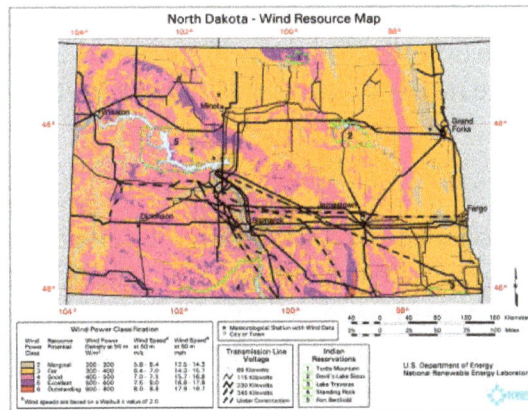

Wind resource map for the windiest U.S. state, North Dakota

Government agencies in some countries publish maps (commonly collected together as a national 'wind atlas') of estimated wind resources, which serve to inform policy-making and encourage wind power development. Examples include the Canadian Wind Atlas, the European Wind Atlas, and the Wind Resource Atlas of the United States. Recognizing the lack of knowledge of wind (and solar) resource potential in developing countries, the Solar and Wind Energy Resource Assessment (SWERA) project was initiated by the United Nations Environment Program in 2002, with funding from the Global Environment Facility, to carry out initial mapping using only satellite-based data. More recently the Energy Sector Management Assistance Program (ESMAP), a program within the World Bank, has launched an initiative to map wind and other renewable energy resources in a number of developing countries, with the intention of developing high quality mapping outputs (and associated datasets) that are validated with specially commissioned ground-based data. There is also an ongoing effort by the International Renewable Energy Agency (IRENA) to create a Global Atlas for Renewable Energy, which brings together publicly available GIS data on wind and other renewable energy resources.

Wind prospecting can begin with the use of such maps, but the lack of accuracy and fine detail make them useful only for preliminary selection of sites for collecting wind speed data. With increasing numbers of ground-based measurements from specially installed anemometer stations, as well as operating data from commissioned wind farms, the accuracy of wind resource maps in many countries has improved over time, although coverage in most developing countries is still patchy. In addition to the publicly available sources listed above, maps are available as commercial products through specialist consultancies, or users of GIS software can make their own using publicly available GIS data such as the US National Renewable Energy Laboratory's High Resolution Wind Data Set.

Although the accuracy has improved, it is unlikely that wind resource maps, whether public or commercial, will eliminate the need for on-site measurements for utility-scale wind generation projects. However, mapping can help speed up the process of site identification and the existence of high quality, ground-based data can shorten the amount of time that on-site measurements need to be collected.

Measurements

To estimate the energy production of a wind farm, developers must first measure the wind on site. Meteorological towers equipped with anemometers, wind vanes, and sometimes temperature, pressure, and relative humidity sensors are installed. Data from these towers must be recorded for at least one year to calculate an annually representative wind speed frequency distribution.

Since onsite measurements are usually only available for a short period, data is also collected from nearby long-term reference stations (usually at airports). This data is used to adjust the onsite measured data so that the mean wind speeds are representative of a long-term period for which onsite measurements are not available. Versions of these maps can be seen and used with software applications such as windNavigator.

Calculations

The following calculations are needed to accurately estimate the energy production of a proposed wind farm project:

- Correlations between onsite meteorological towers:
    - Multiple meteorological towers are usually installed on large wind farm

sites. For each tower, there will be periods of time where data is missing but has been recorded at another onsite tower. Least squares linear regressions can be used to fill in the missing data. These correlations are more accurate if the towers are located near each other (a few km distance), the sensors on the different towers are of the same type, and are mounted at the same height above the ground.

- Correlations between long term weather stations and onsite meteorological towers:

  - Because wind is highly variable year to year, short-term (< 5 years) onsite measurements can result in highly inaccurate energy estimates. Therefore, wind speed data from nearby longer term weather stations (usually located at airports) are used to adjust the onsite data. Least squares linear regressions are usually used, although several other methods exist as well.

- Vertical shear to extrapolate measured wind speeds to turbine hub height:

  - The hub heights of modern wind turbines are usually 80 m or greater, but cost effective meteorological towers are only available up to 60 m in height. The power law and log law vertical shear profiles are the most common methods of extrapolating measured wind speed to hub height.

- Wind flow modeling to extrapolate wind speeds across a site:

  - Wind speeds can vary considerably across a wind farm site if the terrain is *complex* (hilly) or there are changes in *roughness* (the height of vegetation or buildings). Wind flow modeling software, based on either the traditional WAsP linear approach or the newer CFD approach, is used to calculate these variations in wind speed.

- Energy production using a wind turbine manufacturer's power curve:

  - When the long term hub height wind speeds have been calculated, the manufacturer's power curve is used to calculate the gross electrical energy production of each turbine in the wind farm.

- Application of energy loss factors:

  - To calculate the net energy production of a wind farm, the following loss factors are applied to the gross energy production:

    - wind turbine wake loss
    - wind turbine availability
    - electrical losses

- blade degradation from ice/dirt/insects

- high/low temperature shutdown

- high wind speed shutdown

- curtailments due to grid issues

Software Applications

Wind power developers use various types of software applications to assess wind resources.

Wind Data Management

Wind data management software assists the user in gathering, storing, retrieving, analyzing, and validating wind data. Typically the wind data sets are collected directly from a data logger, located at a meteorological monitoring site, and are imported into a database. Once the data set is in the database it can be analyzed and validated using tools built into the system or it can be exported for use in external wind data analysis software, wind flow modeling software, or wind farm modeling software.

Many data logger manufacturers offer wind data management software that is compatible with their logger. These software packages will typically only gather, store, and analyze data from the manufacturer's own loggers.

Third party data management software and services exist that can accept data from a wide variety of loggers and offer more comprehensive analysis tools and data validation.

Wind Data Analysis

Wind data analysis software assist the user in removing measurement errors from wind data sets and perform specialized statistical analysis.

Atmospheric Simulation Modeling

Wind flow modeling methods, described in the following section, named 'Wind flow modeling', provide insights into very high-resolution wind flow behavior, often, at horizontal resolution finer than 100-m. Because of such finest resolution computational fluid dynamics (CFD) modeling application, the typical model domains used by these small-scale models have a few kilometers in the horizontal direction and several hundred meters in the vertical direction. The above-mentioned model domain limitations by small-scale CFD models, need to be addressed and are often addressed by atmospheric CFD models, that could cover horizontal model domains on the order of hundreds of kilometers and vertical domain depths of tens of kilometers. In other words, any atmospheric processes that occur within such large-scale atmospheric model domains, that will influence site-specific wind and its temporal variation should be captured for successful wind resource assessment. This class of atmospheric CFD models and their contribution has not been fully explored and adopted yet, although the aforementioned atmospheric influences captured by such models are highly relevant and critical for the overall wind resource assessment efforts. There are a few such atmospheric CFD models being applied for wind resource assessments today.

Wind Flow Modeling

Wind flow modeling software aims to predict important characteristics of the wind resource at locations where measurements are not available. The most commonly used such software application is WAsP, created at Risø National Laboratory in Denmark. WAsP uses a potential flow model to predict how wind flows over the terrain at a site. WindSim is a similar application that uses CFD calculations instead, which are potentially more accurate, particularly for complex terrains. Fluidyn PANEOLE is another software based on CFD, which makes it a high precision tool capable of generating wind field atlas for wind farm siting, while also integrating wake effects. It generates a wind field atlas for efficient positioning of turbines. Fluidyn PANEOLE also includes local boundary layer effects such as flow detachment, venturi effect between hills or large buildings, surface roughness generated turbulence or sea breeze.

Wind Farm Modeling

Wind farm modeling software aims to simulate the behavior of a proposed or existing wind farm, most importantly to calculate its energy production. The user can usually input wind data, height and roughness contour lines, wind turbine specifications, background maps, and define objects that represent environmental restrictions. This information is then used to design a wind farm that maximizes energy production while taking restrictions and construction issues into account. There are several wind farm modeling software applications available, including Openwind, Windfarmer, WindPRO, WindSim, Fluidyn PANEOLE, meteodyn WT and WAsP.

Medium Scale Wind Farm Modelling

In recent years a new breed of wind farm development has grown from the increased need for distributed generation of electricity from local wind resources. This type of wind projects is mostly driven by land owners with high energetic requirements such as farmers and industrial site managers. A particular requirement from a wind modelling point of view is the inclusion of all local features such as trees, hedges and buildings as turbine hub-heights range from as little as 10m to 50m. Wind modelling approaches need to include these features but very few of the available wind modelling commercial software provide this capability. Several work groups have been set up around the world to look into this modelling requirement and companies including Digital Engineering Ltd (UK), NREL (USA), DTU Wind Energy (Denmark) are at the forefront of development in this area and look at the application of meso-CFD wind modelling techniques for this purpose.

References

- M. Lange and U. Focken. Physical approach to short-term wind power forecast, Springer, ISBN 3-540-25662-8, 2005

- Erich Hau (2006). Wind turbines: fundamentals, technologies, application, economics. Birkhäuser. pp. 568, 569. ISBN 978-3-540-24240-6. Retrieved 17 April 2011.

- Sio-Iong Ao; Len Gelman (29 June 2011). Electrical Engineering and Applied Computing. Springer. p. 41. ISBN 978-94-007-1191-4. Retrieved 15 July 2011.

- "Overview of Compressed Air Energy Storage" (PDF). Boise State University. p. 2. Retrieved 2011-07-15.

- "Proposals for Ladymoor Renewable Energy Project" Renew ND. Retrieved 2 November 2007 Archived 18 July 2011 at the Wayback Machine.

- "Stochastic Joint Optimization of Wind Generation and Pumped-Storage Units in an Electricity Market". IEEE. 22 April 2008. Retrieved 2011-04-14.

- Bonnier Corporation (April 1983). Popular Science. Bonnier Corporation. pp. 85, 86. ISSN 0161-7370. Retrieved 17 April 2011.

- "Feasibility Study of Pumped Hydro Energy Storage for Ramea Wind-Diesel Hybrid Power System" (PDF). Memorial University of Newfoundland. Retrieved 2011-04-17.

- "Final Report: Lower Brule Sioux Tribe Wind-Pumped Storage Feasibility Study Project" (PDF). United States Department of Energy. Retrieved 2011-04-17.

- "WHL Energy Limited (WHL)" is an Australian publicly listed company focused on developing and commercializing energy assets including wind energy, solar, biomass and clean fossil fuels. Retrieved 4 July 2010.

Impact of Wind Power on the Environment

Wind power generates the least global warming as per unit of electrical energy produced. Wind power consumes less land, generates less greenhouse gas and is also renewable. In comparison the effect wind power has to the effect is relatively minor to the effect fossil fuels have to the environment. This section explains to the reader the importance of wind power in contemporary times.

Livestock grazing near a wind turbine.

The environmental impact of wind power when compared to the environmental impacts of fossil fuels, is relatively minor. Compared with other low carbon power sources, wind turbines have some of the lowest global warming potential per unit of electrical energy generated. According to the IPCC, in assessments of the life-cycle global warming potential of energy sources, wind turbines have a median value of between 12 and 11 (gCO_{2eq}/kWh) depending on whether off- or onshore turbines are being assessed.

While a wind farm may cover a large area of land, many land uses such as farming and grazing are compatible with it, as only small areas of turbine foundations and infrastructure are made unavailable for use.

Wind turbines generate some noise. At a residential distance of 300 metres (980 ft) this may be around 45 dB, which is slightly louder than a refrigerator. At 1.5 km (1 mi) distance they become inaudible. There are anecdotal reports of negative health effects from noise on people who live very close to wind turbines. Peer-reviewed research has generally not supported these claims. Scientific evidence suggests that wind turbines, when properly sited, do not contribute to health problems.

Aesthetic aspects of wind turbines and resulting changes of the visual landscape can be significant. Conflicts arise especially in scenic and heritage protected landscapes. Siting restrictions (such as setbacks) have often been implemented to limit any intrusive environmental impacts.

There are reports of bird and bat mortality at wind turbines as there are around other artificial structures. The scale of the ecological impact may or may not be significant, depending on specific circumstances. Prevention and mitigation of wildlife fatalities, and protection of peat bogs, affect the siting and operation of wind turbines.

Basic Operational Considerations

Net Energy Gain

Modern wind turbine systems have a net energy gain, in other words during their service life they produce more energy than is used to build the system. Any practical large-scale energy source must produce more energy than is used in its construction. The energy return on investment (EROI) for wind energy is equal to the cumulative electricity generated divided by the cumulative primary energy required to build and maintain a turbine. According to a meta study, in which all existing studies from 1977 to 2007 were reviewed, the EROI for wind ranges from 5 to 35, with the most common turbines in the range of 2 MW nameplate capacity-rotor diameters of 66 meters, the EROI is on average 16. EROI is strongly proportional to turbine size, and larger late-generation turbines average at the high end of this range, at or above 35. Since energy produced is several times energy consumed in construction, there is a net energy gain. Wind turbine manufacturer Vestas claims that initial energy "pay back" is within about 7–9 months of operation for a 1.65-2.0MW wind power plant under low wind conditions, whereas Siemens Wind Power calculates 5–10 months depending on circumstances.

Pollution & Effects on the Grid

Pollution Costs

Wind power consumes no water for continuing operation, and has near negligible emissions directly related to its electricity production. In full life cycle assessments(LCAs), Wind turbines when isolated from the electric grid produce negligible amounts of carbon dioxide, carbon monoxide, sulfur dioxide, nitrogen dioxide, mercury and radioactive waste when in operation, unlike fossil fuel sources and nuclear energy station fuel production, respectively. However, while they produce none in operation, during construction, wind turbines do produce slightly more particulate matter(PM), a form of air pollution, at a rate per unit of energy generated(kWh) higher than a fossil gas electricity station("NGCC"), and also more heavy metals and PM than nuclear stations per unit of energy generated. As far as total pollution costs in economic terms, in a comprehensive 2006 European study, alpine Hydropower was found to exhibit the lowest external pollution, or externality, costs of all electricity generating systems, below 0.05 c€/kWh. Wind power externality costs were found to be 0.09 - 0.12c€/kW, while nuclear energy had a 0.19 c€/kWh value and fossil fuels generated 1.6 - 5.8 c€/kWh of downstream costs. With the exception of the latter fossil fuels, these are negligible costs in comparison to the cost of electricity production, which is approximately 10 c€/kWh in European countries.

Findings when Connected to the Grid

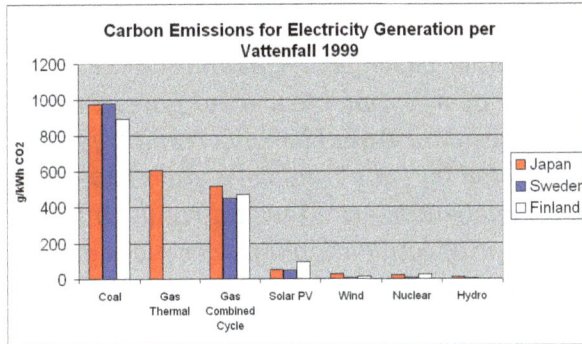

The Vattenfall utility company study found Hydroelectric, nuclear stations and wind turbines to have far less greenhouse emissions than other sources represented.

A typical study of a wind farms Life cycle assessment, when not connected to the electric grid, usually results in similar findings as the following 2006 analysis of 3 installations in the US Midwest, where the carbon dioxide(co2) emissions of wind power ranged from 14 to 33 tonnes (15 to 36 short tons) per GWh(14 - 33 gCO_2/kWh) of energy produced, with most of the CO_2 emission intensity coming from producing the concrete for wind-turbine foundations. By combining similar data from numerous individual studies in a meta-analysis, the median global warming potential for wind power was found to be 11-12g CO2/kWh and unlikely to change significantly.

However these relatively low pollution values begin to increase as greater and greater wind energy is added to the grid, or wind power 'electric grid penetration' levels are reached. Due to the effects of attempting to balance out the energy demands on the grid, from Intermittent power sources e.g. wind power(sources which have low capacity factors due to the weather), this either requires the construction of large energy storage projects, which have their own emission intensity which must be added to wind power's system-wide pollution effects, or it requires more frequent reliance on fossil fuels than the spinning reserve requirements necessary to back up more dependable sources. With the latter combination presently being the more common.

This higher dependence on back-up/Load following power plants to ensure a steady power grid output has the knock-on-effect of more frequent inefficient(in CO$_2$e g/kWh) throttling up and down of these other power sources in the grid to facilitate the intermittent power source's variable output. When one includes the total effect of intermittent sources on other power sources in the grid system, that is, including these inefficient start up emissions of backup power sources to cater for wind energy, into wind energy's total system wide life cycle, this results in a higher real-world wind energy emission intensity. Higher than the direct g/kWh value that is determined from looking at the power source in isolation and thus ignores all down-stream detrimental/inefficiency effects it has on the grid. This higher dependence on back-up/Load following power plants to ensure a steady power grid output forces fossil power plants to operate in less efficient states. In a 2012 paper that appeared in the *Journal of Industrial Ecology* it states.

"The thermal efficiency of fossil-based power plants is reduced when operated at fluctuating and suboptimal loads to supplement wind power, which may degrade, to a certain extent, the GHG(Greenhouse gas) benefits resulting from the addition of wind to the grid. A study conducted by Pehnt and colleagues (2008) reports that a moderate level of [grid] wind penetration (12%) would result in efficiency penalties of 3% to 8%, depending on the type of conventional power plant considered. Gross and colleagues (2006) report similar results, with efficiency penalties ranging from nearly 0% to 7% for up to 20% [of grid] wind penetration. Pehnt and colleagues (2008) conclude that the results of adding offshore wind power in Germany on the background power systems maintaining a level supply to the grid and providing enough reserve capacity amount to adding between 20 and 80 g CO2-eq/kWh to the life cycle GHG emissions profile of wind power."'

In comparison to other low carbon power sources Wind turbines, when assessed in isolation, have a median life cycle emission value of between 12 and 11 (gCO$_{2eq}$/kWh). While the more dependable alpine Hydropower and nuclear stations have median total life cycle emission values of 24 and 12 g CO2-eq/kWh respectively.

While an increase in emissions due to grid connection is an issue, Pehnt et al. still conclude that these 20 and 80 g CO2-eq/kWh added penalties still result in wind being roughly ten times less polluting than fossil gas and coal which emit ~400 and 900 g CO2-eq/kWh respectively.

However, as these losses occur due to cycling of fossil power plants, they become smaller when more renewables are added to the power system and fossil power plants are replaced by renewables. Consequently, also the emissions of the renewables drop with the progressing energy transition.

Rare-earth Use

The production of permanent magnets used in some wind turbines makes use of neodymium. Primarily exported by China, pollution concerns associated with the extraction of this rare-earth element have prompted government action in recent years, and international research attempts to refine the extraction process. Research is underway on turbine and generator designs which reduce the need for neodymium, or eliminate the use of rare-earth metals altogether. Additionally, the large wind turbine manufacturer Enercon GmbH chose very early not to use permanent magnets for its direct drive turbines, in order to avoid responsibility for the adverse environmental impact of rare earth mining.

Ecology

Land Use

Wind farms are often built on land that has already been impacted by land clearing. The vegetation clearing and ground disturbance required for wind farms is minimal compared with coal mines and coal-fired power stations. If wind farms are decommissioned, the landscape can be returned to its previous condition.

A study by the US National Renewable Energy Laboratory of US wind farms built between 2000 and 2009 found that, on average, only 1.1 percent of the total wind farm area suffered surface disturbance, and only 0.43 percent was permanently disturbed by wind power installations. On average, there were 63 hectares (156 acres) of total wind farm area per MW of capacity, but only 0.27 hectares (0.67 acres) of permanently disturbed area per MW of wind power capacity.

In the UK many prime wind farm sites - locations with the best average wind speeds - are in upland areas which are frequently covered by blanket bog. This type of habitat exists in areas of relatively high rainfall where large areas of land remain permanently sodden. Construction work may create a risk of disruption to peatland hydrology which could cause localised areas of peat within the area of a wind farm to dry out, disintegrate, and so release their stored carbon. At the same time, the warming climate which renewable energy schemes seek to mitigate could itself pose an existential threat to peatlands throughout the UK. A Scottish MEP campaigned for a moratorium on wind developments on peatlands saying that "Damaging the peat causes the release of more carbon dioxide than wind farms save". A 2014 report for the Northern Ireland Environment Agency noted that siting wind turbines on peatland could release considerable carbon dioxide from the peat, and also damage the peatland contributions to flood control and water quality: "The potential knock-on effects of using the peatland resource for wind turbines are considerable and it is arguable that the impacts on this facet of biodiversity will have the most noticeable and greatest financial implications for Northern Ireland."

Farmers and graziers often lease land to companies building wind farms. In the U.S., landowners may receive annual lease payments of two thousand to five thousand dollars per turbine. In 2011, taxes, fees, and assessments of wind power facilities enabled Sherman County, Oregon to pay each landowner $590 per year "...to reward residents who have made no financial gains [directly] from wind energy development, but whose views of... [the] landscape now include a panorama of turbines".

Wind-energy advocates contend that less than 1% of the land is used for foundations and access roads, the other 99% can still be used for farming. A wind turbine needs about 200–400 m² for the foundation. A (small) 500-kW-turbine with an annual production of 1.4 GWh produces 11.7 MWh/m², which is comparable with coal-fired plants (about 15-20 MWh/m²), coal-mining not included. With increasing size of the wind turbine the relative size of the foundation decreases. Critics point out that on some locations in forests the clearing of trees around tower bases may be necessary for installation sites on mountain ridges, such as in the northeastern U.S. This usually takes the clearing of 5,000 m² per wind turbine.

Turbines are not generally installed in urban areas. Buildings interfere with wind, turbines must be sited a safe distance ("setback") from residences in case of failure, and the value of land is high. There are a few notable exceptions to this. The WindShare ExPlace wind turbine was erected in December 2002, on the grounds of Exhibition Place, in Toronto, Canada. It was the first wind turbine installed in a major North American urban city centre. Steel Winds also has a 20 MW urban project south of Buffalo, New York. Both of these projects are in urban locations, but benefit from being on uninhabited lake shore property.

Livestock

The land can still be used for farming and cattle grazing. Livestock are unaffected by the presence of wind farms. International experience shows that livestock will "graze right up to the base of wind turbines and often use them as rubbing posts or for shade".

In 2014, a first of its kind Veterinary study attempted to determine the effects of rearing livestock near a wind turbine, the study compared the health effects of a wind turbine on the development of two groups of growing geese, preliminary results found that geese raised within 50 meters of a wind turbine gained less weight and had a higher concentration of the stress hormone cortisol in their blood than geese at a distance of 500 meters.

Semi-domestic reindeer avoid the construction activity, but seem unaffected when the turbines are operating.

Impact on Wildlife

Environmental assessments are routinely carried out for wind farm proposals, and potential impacts on the local environment (e.g. plants, animals, soils) are evaluated. Turbine locations and operations are often modified as part of the approval process to avoid or minimise impacts on threatened species and their habitats. Any unavoidable impacts can be offset with conservation improvements of similar ecosystems which are unaffected by the proposal.

A research agenda from a coalition of researchers from universities, industry, and government, supported by the Atkinson Center for a Sustainable Future, suggests modeling the spatiotemporal patterns of migratory and residential wildlife with respect to geographic features and weather, to provide a basis for science-based decisions about where to site new wind projects. More specifically, it suggests:

- Use existing data on migratory and other movements of wildlife to develop predictive models of risk.

- Use new and emerging technologies, including radar, acoustics, and thermal imaging, to fill gaps in knowledge of wildlife movements.

- Identify specific species or sets of species most at risk in areas of high potential wind resoures.

Birds

Data largely from a preliminary study, conducted by B. Sovacool, into causes of avian mortality in the United States, annual		
Source	Estimated mortality (in millions)	Estimated deaths (per GWh)
Wind turbines	0.02 – 0.57	0.269
Aircraft	0.08	(n/a)
Nuclear power plants	0.33 – 0	0.416 – 0
Oilfield oil waste & waste water pits	0.50 – 1	(n/a)
Nuisance bird control kills (airports, agriculture, etc...)	2	(n/a)
Communication towers (cellular, radio, microwave)	4 – 50	(n/a)
Large communications towers (over 180', N. America)	6.8	(n/a)
Fossil fuel powerplants	14	5.18
Cars & trucks	50 – 100	(n/a)
Agriculture	67	(n/a)
Pesticide use	72	(n/a)
Hunting	100 – 120	(n/a)
Transmission lines (conventional powerplants)	174 – 175	(n/a)
Buildings and windows	365 – 988	(n/a)
Domestic and feral cats	210 – 3,700	(n/a)

The impact of wind energy on birds, which can fly into turbines directly, or indirectly have their habitats degraded by wind development, is complex. Projects such as the Black Law Wind Farm have received wide recognition for its contribution to environmental objectives, including praise from the Royal Society for the Protection of Birds, who describe the scheme as both improving the landscape of a derelict opencast mining site and also benefiting a range of wildlife in the area, with an extensive habitat management projects covering over 14 square kilometres.

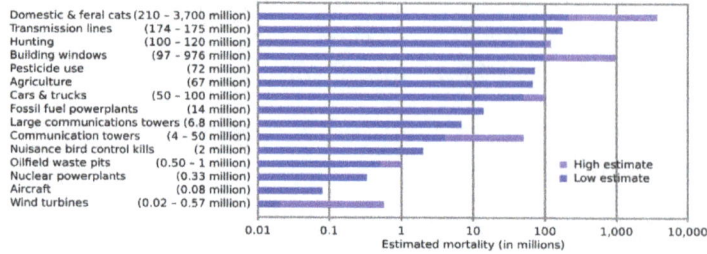

The preliminary data, from the above table during 2013, 'Causes of avian mortality in the United States, annual', shown as a bar graph, inclusive of a high nuclear-fission bird mortality figure that the author later recognized was due to a major error on their part.

The meta-analysis on avian mortality by Benjamin K. Sovacool led him to suggest that there were a number of deficiencies in other researchers' methodologies. Among them, he stated were a focus on bird deaths, but not on the reductions in bird births: for example, mining activities for fossil fuels and pollution from fossil fuel plants have led to significant toxic deposits and acid rain that have damaged or poisoned many nesting and feeding grounds, leading to reductions in births. The large cumulated footprint of wind turbines, which reduces the area available to wildlife or agriculture, is also missing from all studies including Sovacool's. Many of the studies also made no mention of avian deaths per unit of electricity produced, which excluded meaningful comparisons between different energy sources. More importantly, it concluded, the most visible impacts of a technology, as measured by media exposure, are not necessarily the most flagrant ones.

Sovacool estimated that in the United States wind turbines kill between 20,000 and 573,000 birds per year, and although he regards either figure is minimal compared to bird deaths from other causes. He uses the lower 20,000 figure in his study and table to arrive at a direct mortality rate per unit of energy generated figure of 0.269 per GWh for wind power. Fossil-fueled power plants, which wind turbines generally require to make up for their weather dependent intermittency, kill almost 20 times as many birds per gigawatt hour (GWh) of electricity according to Sovacool. Bird deaths due to other human activities and cats total between 797 million and 5.29 billion per year in the U.S. Additionally, while many studies concentrate on the analysis of bird deaths, few have been conducted on the reductions of bird births, which are the additional consequences of the various pollution sources that wind power partially mitigates. Of the bird deaths Sovacool attributed to fossil-fuel power plants, 96 percent were due to the effects of climate change. While the study did not assess bat mortality due to various forms of energy, he considered it not unreasonable to assume a similar ratio of mortality. The Sovacool study has provoked controversy because of its treatment of data. In a series of replies, Sovacool acknowledged a number of large errors, particularly those that relate to his earlier "0.33 to 0.416" fatalities overestimate for the number of bird deaths per GWh of nuclear power, and cautioned that "the study already tells you the numbers are very rough estimates that need to be improved."

A 2013 meta-analysis by Smallwood identified a number of factors which result in serious under-reporting of bird and bat deaths by wind turbines. These include inefficient searches, inadequate search radius, and carcass removal by predators. To adjust the results of different studies, he applied correction factors from hundreds of carcass placement trials. His meta-analysis concluded that in 2012 in the United States, wind turbines resulted in the deaths of 888,000 bats and 573,000 birds, including 83,000 birds of prey.

Also in 2013, a meta-analysis by SCott Loss and others in the journal *Biological Conservation* found that the likely mean number of birds killed annually in the U.S by monopole tower wind turbines was 234,000. The authors acknowledged the larger number reported by Smallwood, but noted that Smallwood's meta-analysis did not distinguish between types of wind turbine towers. The monopole towers used almost exclusively for new wind installations have mortality rates that "increase with increasing height of monopole turbines", but as of yet, it remains to be determined if increasingly taller monopole towers result in lower mortality per GWh.

Bird mortality at wind energy facilities can vary greatly depending on the location, construction, and height, with some facilities reporting zero bird fatalities, and others as high as 9.33 birds per turbine per year. A 2007 article in the journal *Nature* stated that each wind turbine in the U.S. kills an average of 0.03 birds per year, and recommends that more research needs to be done.

A comprehensive study of wind turbine bird deaths by the Canadian Wildlife Service in 2013 analyzed reports from 43 out of the 135 wind farms operating across Canada as of December 2011. After adjusting for search inefficiencies, the study found an average of 8.2 bird deaths per tower per year, from which they arrived at a total of 23,000 per year for Canada at that time. Actual habitat loss averaged 1.23 hectares per turbine, which involved the direct loss of, on average, 1.9 nesting sites per turbine. The effective habitat loss, which was not quantified, was observed to be highly variable between species: some species avoided nesting within 100 to 200 m from turbines, while other species were observed feeding on the ground directly under the blades. The study concluded that, overall, the combined effect on birds was "relatively small" compared to other causes of bird mortality, but noted that mitigation measures might be required in some situations to protect at-risk species.

Wind facilities have attracted the most attention for impacts on iconic raptor species, including golden eagles. The Pine Tree Wind energy project near Tehachapi, California has one of the highest raptor mortality rates in the country; by 2012 at least eight golden eagles had been killed according to the U.S. Fish and Wildlife Service (USFWS). Biologists have noted that it is more important to avoid losses of large birds as they have lower breeding rates and can be more severely impacted by wind turbines in certain areas.

Large numbers of bird deaths are also attributed to collisions with buildings. An estimated 1 to 9 million birds are killed every year by tall buildings in Toronto, Canada alone, according to the wildlife conservation organization Fatal Light Awareness Program. Other studies have stated that 57 million are killed by cars, and some 365 to 988 million are killed by collisions with buildings and plate glass in the United States alone. Promotional event lightbeams as well as ceilometers used at airport weather offices can be particularly deadly for birds, as birds become caught in their lightbeams and suffer exhaustion and collisions with other birds. In the worst recorded ceilometer lightbeam kill-off during one night in 1954, approximately 50,000 birds from 53 different species died at the Warner Robins Air Force Base in the United States.

Arctic terns and a wind turbine at the Eider Barrage in Germany.

In the United Kingdom, the Royal Society for the Protection of Birds (RSPB) concluded that "The available evidence suggests that appropriately positioned wind farms do not pose a significant hazard for birds." It notes that climate change poses a much more significant threat to wildlife, and therefore supports wind farms and other forms of renewable energy as a way to mitigate future damage. In 2009 the RSPB warned that "numbers of several breeding birds of high conservation concern are reduced close to wind turbines" probably because "birds may use areas close to the turbines less often than would be expected, potentially reducing the wildlife carrying capacity of an area.

Concerns have been expressed that wind turbines at Smøla, Norway are having a deleterious effect on the population of white-tailed eagles, Europe's largest bird of prey. They have been the subject of an extensive re-introduction programme in Scotland, which could be jeopardised by the expansion of wind turbines.

The Peñascal Wind Power Project in Texas is located in the middle of a major bird migration route, and the wind farm uses avian radar originally developed for NASA and the United States Air Force to detect birds as far as 4 miles (6.4 km) away. If the system determines that the birds are in danger of running into the rotating blades, the turbines shut down and are restarted when the birds have passed. A 2005 Danish study used surveillance radar to track migrating birds traveling around and through an offshore wind farm. Less than 1% of migrating birds passing through an offshore wind farm in Rønde, Denmark, got close enough to be at risk of collision, though the site was studied only during low-wind conditions. The study suggests that migrating birds may avoid large turbines, at least in the low-wind conditions the research was conducted in.

Old style wind turbines at Altamont Pass in California, which are being replaced by more "bird-friendly designs". While newer designs are taller, there is as yet, no definitive evidence that they are "friendlier". A recent study suggests that they might not be safer to wildlife, and are not a "simple fix", according to Oklahoma State University ecologist Scott Loss.

In 2012, researchers reported that, based on their four-year radar tracking study of birds after construction of an offshore wind farm near Lincolnshire, that pink-footed geese migrating to the U.K. to overwinter altered their flight path to avoid the turbines.

At the Altamont Pass Wind Farm in California, a settlement between the Audubon Society, Californians for Renewable Energy and NextEra Energy Resources who operate some 5,000 turbines in the area requires the latter to replace nearly half of the smaller turbines with newer, more bird-friendly models by 2015 and provide $2.5 million for raptor habitat restoration. The proposed Chokecherry and Sierra Madre Wind project in Wyoming, however, is expected to kill nearly 5,400 birds each year, including over 150 raptors, according to a Bureau of Land Management environmental analysis.

Bats

Bats may be injured by direct impact with turbine blades, towers, or transmission lines. Recent research shows that bats may also be killed when suddenly passing through a low air pressure region surrounding the turbine blade tips.

The numbers of bats killed by existing onshore and near-shore facilities have troubled bat enthusiasts.

In April 2009 the Bats and Wind Energy Cooperative released initial study results showing a 73% drop in bat fatalities when wind farm operations are stopped during low wind conditions, when bats are most active. Bats avoid radar transmitters, and placing microwave transmitters on wind turbine towers may reduce the number of bat collisions.

A 2013 study produced an estimate that wind turbines killed more than 600,000 bats in the U.S. the previous year, with the greatest mortality occurring in the Appalachian Mountains. Some earlier studies had produced estimates of between 33,000 and 888,000 bat deaths per year.

Weather and Climate Change

Wind farms may affect weather in their immediate vicinity. This turbulence from spinning wind turbine rotors increases vertical mixing of heat and water vapor that affects the meteorological conditions downwind, including rainfall. Overall, wind farms lead to a slight warming at night and a slight cooling during the day time. This effect can be reduced by using more efficient rotors or placing wind farms in regions with high natural turbulence. Warming at night could "benefit agriculture by decreasing frost damage and extending the growing season. Many farmers already do this with air circulators".

A number of studies have used climate models to study the effect of extremely large wind farms. One study reports simulations that show detectable changes in global climate for very high wind farm usage, on the order of 10% of the world's land area. Wind power has a negligible effect on global mean surface temperature, and it would deliver "enormous global benefits by reducing emissions of CO_2 and air pollutants". Another peer-reviewed study suggested that using wind turbines to meet 10 percent of global energy demand in 2100 could actually have a warming effect, causing temperatures to rise by 1 °C (1.8 °F) in the regions on land where the wind farms are installed, including a smaller increase in areas beyond those regions. This is due to the effect of wind turbines on both horizontal and vertical atmospheric circulation. Whilst turbines installed in water would have a cooling effect, the net impact on global surface temperatures would be an increase of 0.15 °C (0.27 °F). Author Ron Prinn cautioned against interpreting the study "as an argument against wind power, urging that it be used to guide future research". "We're not pessimistic about wind," he said. "We haven't absolutely proven this effect, and we'd rather see that people do further research".

Impacts on People

Aesthetics

The surroundings of Mont Saint-Michel at low tide. While windy coasts are good locations for wind farms, aesthetic considerations may preclude such developments in order to preserve historic views of cultural sites.

Aesthetic considerations of wind power stations have often a significant role in their evaluation process. To some, the perceived aesthetic aspects of wind power stations may conflict with the protection of historical sites. Wind power stations are less likely to be perceived negatively in urbanized and industrial regions. Aesthetic issues are subjective and some people find wind farms pleasant or see them as symbols of energy independence and local prosperity. While studies in Scotland predict wind farms will damage tourism, in other countries some wind farms have themselves become tourist attractions, with several having visitor centers at ground level or even observation decks atop turbine towers.

In the 1980s, wind energy was being discussed as part of a soft energy path. Renewable energy commercialization led to an increasing industrial image of wind power, which is being criticized by various stakeholders in the planning process, including nature protection associations. Newer wind farms have larger, more widely spaced turbines, and have a less cluttered appearance than older installations. Wind farms are often built on land that has already been impacted by land clearing and they coexist easily with other land uses.

Coastal areas and areas of higher altitude such as ridgelines are considered prime for wind farms, due to constant wind speeds. However, both locations tend to be areas of high visual impact and can be a contributing factor in local communities' resistance to some projects. Both the proximity to densely populated areas and the necessary wind speeds make coastal locations ideal for wind farms.

Loreley rock in Rhineland-Palatinate, part of UNESCO World heritage site Rhine Gorge

Wind power stations can impact on important sight relations which are a key part of culturally important landscapes, such as in the Rhine Gorge or Moselle valley. Conflicts between heritage status of certain areas and wind power projects have arisen in various countries. In 2011 UNESCO raised concerns regarding a proposed wind farm 17 kilometres away from the French island abbey of Mont-Saint-Michel. In Germany, the impact of wind farms on valuable cultural landscapes has implications on zoning and land-use planning. For example, sensitive parts of the Moselle valley and the background of the Hambach Castle, according to the plans of the state government, will be kept free of wind turbines.

Wind turbines require aircraft warning lights, which may create light pollution. Complaints about these lights have caused the US FAA to consider allowing fewer lights per turbine in certain areas. Residents near turbines may complain of "shadow flicker" caused by rotating turbine blades, when the sun passes behind the turbine. This can be avoided by locating the wind farm to avoid unacceptable shadow flicker, or by turning the turbine off for the time of the day when the sun is at the angle that causes flicker. If a turbine is poorly sited and adjacent to many homes, the duration of shadow flicker on a neighbourhood can last hours.

Wind Turbine Syndrome

Wind turbine syndrome is a psychosomatic disorder largely caused by anxiety about wind farms and not by the turbines themselves. There is limited evidence of anxiety effects caused by low level noise in the close vicinity of the turbines.

Safety

Some turbine nacelle fires cannot be extinguished because of their height, and are sometimes left to burn themselves out. In such cases they generate toxic fumes and can cause secondary fires below. However, newer wind turbines are built with automatic fire extinguishing systems similar to those provided for jet aircraft engines. These autonomous systems, which can be retrofitted to older wind turbines, automatically detect a fire, order the shut down of the turbine unit and immediately extinguish the fires completely.

During winter ice may form on turbine blades and subsequently be thrown off during operation. This is a potential safety hazard, and has led to localised shut-downs of turbines. Modern turbines can detect ice formation and excess vibration during operations, and are shut down automatically. Electronic controllers and safety sub-systems monitor many aspects of the turbine, generator, tower, and environment to determine if the turbine is operating in a safe manner within prescribed limits. These systems can temporarily shut down the turbine due to high wind, ice, electrical load imbalance, vibration, and other problems. Recurring or significant problems cause a system lockout and notify an engineer for inspection and repair. In addition, most systems include multiple passive safety systems that stop operation even if the electronic controller fails. A 2007 study noted that no insurance claims had been filed, either in Europe or the US, for injuries from ice falling from wind towers, and that while some fatal accidents have occurred to industry workers, only one wind-tower related fatality was known to occur to a non-industry person: a parachutist.

Offshore

Many offshore wind farms have contributed to electricity needs in Europe and Asia for years, and as of 2014 the first offshore wind farms are under development in U.S. waters. While the offshore wind industry has grown dramatically over the last several decades, especially in Europe, there is still some uncertainty associated with how the construction and operation of these wind farms affect marine animals and the marine environment.

Traditional offshore wind turbines are attached to the seabed in shallower waters within the near-shore marine environment. As offshore wind technologies become more advanced, floating structures have begun to be used in deeper waters where more wind resources exist.

Common environmental concerns associated with offshore wind developments include:

- The risk to seabirds being struck by wind turbine blades or being displaced from critical habitats;

- Underwater noise associated with the installation process of monopile turbines;

- The physical presence of offshore wind farms altering the behavior of marine mammals, fish, and seabirds by reasons of either attraction or avoidance;

- Potential disruption of the near-field and far-field marine environments from large offshore wind projects.

Due to the landscape protection status of large areas of the Wadden Sea, a major World Heritage Site with various national parks (e.g. Lower Saxon Wadden Sea National Park) German offshore installations are mostly restricted on areas outside the territorial waters. Offshore capacity in Germany is therefore way behind the British or Danish near coast installments, which face much lower restrictions.

In January 2009, a comprehensive government environmental study of coastal waters in the United Kingdom concluded that there is scope for between 5,000 and 7,000 offshore wind turbines to be installed without an adverse impact on the marine environment. The study—which forms part of the Department of Energy and Climate Change's Offshore Energy Strategic Environmental Assessment—is based on more than a year's research. It included analysis of seabed geology, as well as surveys of sea birds and marine mammals. There does not seem to have been much consideration however of the likely impact of displacement of fishing activities from traditional fishing grounds.

A study published in 2014 suggests that some seals prefer to hunt near turbines, likely due to the laid stones functioning as artificial reefs which attract invertebrates and fish. However, studies of the impacts of dredging on complex soft sediment communities suggest that the impacts caused by construction of structures such as windfarms may still be discernible up to 10 years after

References

- Bird, David Michael. The Bird Almanac: The Ultimate Guide to Essential Facts and Figures of the World's Birds, Key Porter Books, 1999, ISBN 155263003X, ISBN 978-1552630037.

- Professor Simon Chapman. "Summary of main conclusions reached in 25 reviews of the research literature on wind farms and health" Sydney University School of Public Health, April 2015

- "Life cycle assessment of electricity produced from onshore sited wind power plants based on Vestas V82-1.65 MW turbines" page 4. Vestas, 29 December 2006. Accessed: 27 November 2014.

- Wittrup, Sanne. "6 MW vindmølle betaler sig energimæssigt tilbage 33 gange" English translation Ingeniøren, 26 November 2014. Accessed: 27 November 2014.

- Will Newer Wind Turbines Mean Fewer Bird Deaths? The jury is still out on what works to protect wildlife. By Andrew Curry, for National Geographic. 2014

- Warwicker, Michelle. "Seals 'feed' at offshore wind farms, study shows" BBC, 21 July 2014. Accessed: 22 July 2014. Video of seal path

- Prentice, Colin (19 December 2013). "Climate change poses serious threat to Britain's peat bogs". London: Imperial College London. Retrieved 2013-12-19.

- Sovacool, Benjamin K. (2013). "The avian benefits of wind energy: A 2009 update". Renewable Energy. 49: 19–24. doi:10.1016/j.renene.2012.01.074.

- Smallwood, K. S. (2013). "Comparing bird and bat fatality-rate estimates among North American wind-energy projects". Wildlife Society Bulletin. 37: 19–33. doi:10.1002/wsb.260.

- Contaminant Issues - Oil Field Waste Pits, U.S. Fish & Wildlife Service, U.S. Department of the Interior. Retrieved July 30, 2013.

- U.S. Cats Kill Up To 3.7 Billion Birds, 20.7 Billion Small Mammals Annually, Paris: Agence France-Presse, January 29, 2013. Retrieved from The Globe & Mail website, January 30, 2013.

- Lorenzini, Paul (April 30, 2013). "Nukes kill more birds than wind?". Atomic Insights. Retrieved 26 August 2013.

- K. Shawn Smallwood, "Comparing bird and bat fatality-rate estimates among North American wind-energy projects", Wildlife Society Bulletin, 26 Mar. 2013.

- "Estimates of bird collision mortality at wind facilities in the contiguous United States Scott R. Lossa et. al.". Biological Conservation. 168: 201–209. doi:10.1016/j.biocon.2013.10.007.

- Morin, Monte. 600,000 bats killed at wind energy facilities in 2012, study says, LA Times, November 8, 2013.

- "Tourism blown off course by turbines". Berwickshire: The Berwickshire News. 2013-03-28. Retrieved 2013-10-08.

- Fittkau, Ludger: Ästhetik und Windräder, Neues Gutachten zu "Windenergienutzung und bedeutenden Kulturlandschaften" in Rheinland-Pfalz, Kultur heute, 30 July 2013

Progress of Wind Power

The history of wind power can be traced by the evolution of devices that use this energy. Wind mills, sail ships are important landmarks in human technological history. This chapter provides an overview of the history of wind power and wind power related devices.

Charles Brush's windmill of 1888, used for generating electricity.

Wind power has been used as long as humans have put sails into the wind. For more than two millennia wind-powered machines have ground grain and pumped water. Wind power was widely available and not confined to the banks of fast-flowing streams, or later, requiring sources of fuel. Wind-powered pumps drained the polders of the Netherlands, and in arid regions such as the American mid-west or the Australian outback, wind pumps provided water for live stock and steam engines.

With the development of electric power, wind power found new applications in lighting buildings remote from centrally-generated power. Throughout the 20th century parallel paths developed small wind plants suitable for farms or residences, and larger utility-scale wind generators that could be connected to electricity grids for remote use of power. Today wind powered generators operate in every size range between tiny plants for battery charging at isolated residences, up to near-gigawatt sized offshore wind farms that provide electricity to national electrical networks.

By 2014, over 240,000 commercial-sized wind turbines were operating in the world, producing 4% of the world's electricity.

Antiquity

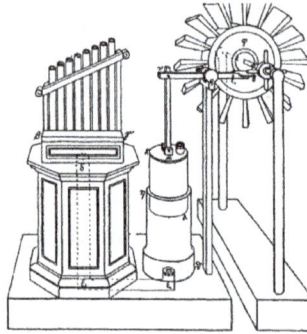

Heron's wind-powered organ, the earliest machine powered by wind

Sailboats and sailing ships have been using wind power for at least 5,500 years, and architects have used wind-driven natural ventilation in buildings since similarly ancient times. The use of wind to provide mechanical power came somewhat later in antiquity.

The Babylonian emperor Hammurabi planned to use wind power for his ambitious irrigation project in the 17th century BC.

The windwheel of the Greek engineer Heron of Alexandria in the 1st century AD is the earliest known instance of using a wind-driven wheel to power a machine. Another early example of a wind-driven wheel was the prayer wheel, which has been used in ancient India, Tibet, and China since the 4th century.

Early Middle Ages

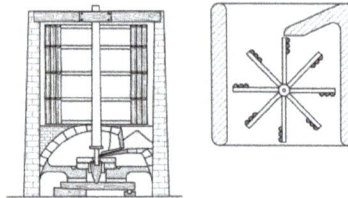

The Persian, horizontal windmill

Medieval depiction of a windmill

The first practical windmills were in use in Sistan, a region in Iran and bordering Afghanistan, at least by the 9th century and possibly as early as the 7th century. These "Panemone windmills" were horizontal windmills,[note 1] which had long vertical drive-shafts with six to twelve rectangular sails covered in reed matting or cloth. These windmills were used to pump water, and in the gristmilling and sugarcane industries. The use of windmills became widespread use across the Middle East and Central Asia, and later spread to China and India. Vertical windmills were later used extensively in Northwestern Europe to grind flour beginning in the 1180s, and many examples still exist. By 1000 AD, windmills were used to pump seawater for salt-making in China and Sicily.

Wind-powered automata are known from the mid-8th century: wind-powered statues that "turned with the wind over the domes of the four gates and the palace complex of the Round City of Baghdad". The "Green Dome of the palace was surmounted by the statue of a horseman carrying a lance that was believed to point toward the enemy. This public spectacle of wind-powered statues had its private counterpart in the 'Abbasid palaces where automata of various types were predominantly displayed."

Late Middle Ages

The vertical windmills of Campo de Criptana were immortalized in chapter VIII of Don Quixote.

The first windmills in Europe appear in sources dating to the twelfth century. These early European windmills were sunk post mills. The earliest certain reference to a windmill dates from 1185, in Weedley, Yorkshire, although a number of earlier but less certainly dated twelfth-century European sources referring to windmills have also been adduced. While it is sometimes argued that crusaders may have been inspired by windmills in the Middle East, this is unlikely since the European vertical windmills were of significantly different design than the horizontal windmills of Afghanistan. Lynn White Jr., a specialist in medieval European technology, asserts that the European windmill was an "independent invention;" he argues that it is unlikely that the Afghanistan-style horizontal windmill had spread as far west as the Levant during the Crusader period. In medieval England rights to waterpower sites were often confined to nobility and clergy, so wind power was an important resource to a new middle class. In addition, windmills, unlike water mills, were not rendered inoperable by the freezing of water in the winter.

By the 14th century Dutch windmills were in use to drain areas of the Rhine River delta.

18th Century

Windmills were used to pump water for salt making on the island of Bermuda, and on Cape Cod during the American revolution. In Mykonos and in other islands of Greece windmills were used to mill flour and remained in use until the early 20th century. Many of them are now refurbished to be inhabited.

19th Century

Blyth's windmill at his cottage in Marykirk in 1891

Wind powered generators were used on ships by the end of the 19th century, as seen on the New Zealand sailing ship "Chance" (1902).

The first windmill used for the production of electricity was built in Scotland in July 1887 by Prof James Blyth of Anderson's College, Glasgow (the precursor of Strathclyde University). Blyth's 10 m high, cloth-sailed wind turbine was installed in the garden of his holiday cottage at Marykirk in Kincardineshire and was used to charge accumulators developed by the Frenchman Camille Alphonse Faure, to power the lighting in the cottage, thus making it the first house in the world to have its electricity supplied by wind power. Blyth offered the surplus electricity to the people of Marykirk for lighting the main street, however, they turned down the offer as they thought electricity was "the work of the devil." Although he later built a wind turbine to supply emergency power to the local Lunatic Asylum, Infirmary and Dispensary of Montrose, the invention never really caught on as the technology was not considered to be economically viable.

Across the Atlantic, in Cleveland, Ohio a larger and heavily engineered machine was designed and constructed in the winter of 1887-1888 by Charles F. Brush, this was built by his engineering company at his home and operated from 1886 until 1900. The Brush wind turbine had a rotor 17 m (56 foot) in diameter and was mounted on an 18 m (60 foot) tower. Although large by today's standards, the machine was only rated at 12 kW; it turned relatively slowly since it had 144 blades. The connected dynamo was used either to charge a bank of batteries or to operate up to 100 incandescent light bulbs, three arc lamps, and various motors in Brush's laboratory. The machine fell into disuse after 1900 when electricity became available from Cleveland's central stations, and was abandoned in 1908.

In 1891 Danish scientist, Poul la Cour, constructed a wind turbine to generate electricity, which was used to produce hydrogen by electrolysis to be stored for use in experiments and to light the Askov High school. He later solved the problem of producing a steady supply of power by inventing a regulator, the Kratostate, and in 1895 converted his windmill into a prototype electrical power plant that was used to light the village of Askov.

In Denmark there were about 2,500 windmills by 1900, used for mechanical loads such as pumps and mills, producing an estimated combined peak power of about 30 MW.

In the American midwest between 1850 and 1900, a large number of small windmills, perhaps six million, were installed on farms to operate irrigation pumps. Firms such as Star, Eclipse, Fairbanks-Morse, and Aeromotor became famed suppliers in North and South America.

20th Century

Development in the 20th century might be usefully divided into the periods:

- 1900–1973, when widespread use of individual wind generators competed against fossil fuel plants and centrally-generated electricity

- 1973–onward, when the oil price crisis spurred investigation of non-petroleum energy sources.

1900–1973

Danish Development

In Denmark wind power was an important part of a decentralized electrification in the first quarter of the 20th century, partly because of Poul la Cour from his first practical development in 1891 at Askov. By 1908 there were 72 wind-driven electric generators from 5 kW to 25 kW. The largest machines were on 24 m (79 ft) towers with four-bladed 23 m (75 ft) diameter rotors. In 1957 Johannes Juul installed a 24 m diameter wind turbine at Gedser, which ran from 1957 until 1967. This was a three-bladed, horizontal-axis, upwind, stall-regulated turbine similar to those now used for commercial wind power development.

Farm Power and Isolated Plants

In 1927 the brothers Joe Jacobs and Marcellus Jacobs opened a factory, Jacobs Wind in Minneapolis to produce wind turbine generators for farm use. These would typically be used for lighting or battery charging, on farms out of reach of central-station electricity and distribution lines. In 30 years the firm produced about 30,000 small wind turbines, some of which ran for many years in remote locations in Africa and on the Richard Evelyn Byrd expedition to Antarctica. Many other manufacturers produced small wind turbine sets for the same market, including companies called Wincharger, Miller Airlite, Universal Aeroelectric, Paris-Dunn, Airline and Winpower.

In 1931 the Darrieus wind turbine was invented, with its vertical axis providing a different mix of design tradeoffs from the conventional horizontal-axis wind turbine. The vertical orientation accepts wind from any direction with no need for adjustments, and the heavy generator and gearbox equipment can rest on the ground instead of atop a tower.

By the 1930s windmills were widely used to generate electricity on farms in the United States where distribution systems had not yet been installed. Used to replenish battery storage banks, these machines typically had generating capacities of a few hundred watts to several kilowatts. Beside providing farm power, they were also used for isolated applications such as electrifying bridge structures to prevent corrosion. In this period, high tensile steel was cheap, and windmills were placed atop prefabricated open steel lattice towers.

The most widely used small wind generator produced for American farms in the 1930s was a two-bladed horizontal-axis machine manufactured by the Wincharger Corporation. It had a peak output of 200 watts. Blade speed was regulated by curved air brakes near the hub that deployed at excessive rotational velocities. These machines were still being manufactured in the United States during the 1980s. In 1936, the U.S. started a rural electrification project that killed the natural market for wind-generated power, since network power distribution provided a farm with more dependable usable energy for a given amount of capital investment.

In Australia, the Dunlite Corporation built hundreds of small wind generators to provide power at isolated postal service stations and farms. These machines were manufactured from 1936 until 1970.

Utility-scale Turbines

A forerunner of modern horizontal-axis utility-scale wind generators was the WIME D-30 in service in Balaklava, near Yalta, USSR from 1931 until 1942. This was a 100 kW generator on a 30 m (100 ft) tower, connected to the local 6.3 kV distribution system. It had a three-bladed 30 metre rotor on a steel lattice tower. It was reported to have an annual load factor of 32 per cent, not much different from current wind machines.

The world's first megawatt-sized wind turbine near Grandpa's Knob Summit,, Castleton, Vermont.

Experimental wind turbine at Nogent-le-Roi, France, 1955.

In 1941 the world's first megawatt-size wind turbine was connected to the local electrical distribution system on the mountain known as Grandpa's Knob in Castleton, Vermont, USA. It was designed by Palmer Cosslett Putnam and manufactured by the S. Morgan Smith Company. This 1.25 MW Smith-Putnam turbine operated for 1100 hours before a blade failed at a known weak point, which had not been reinforced due to war-time material shortages. No similar-sized unit was to repeat this "bold experiment" for about forty years.

Fuel-saving Turbines

During the Second World War, small wind generators were used on German U-boats to recharge submarine batteries as a fuel-conserving measure. In 1946 the lighthouse and residences on the island of Neuwerk were partly powered by an 18 kW wind turbine 15 metres in diameter, to economize on diesel fuel. This installation ran for around 20 years before being replaced by a submarine cable to the mainland.

The Station d'Etude de l'Energie du Vent at Nogent-le-Roi in France operated an experimental 800 KVA wind turbine from 1956 to 1966.

The NASA/DOE 7.5 megawatt Mod-2 three turbine cluster in Goodnoe Hills, Washington in 1981.

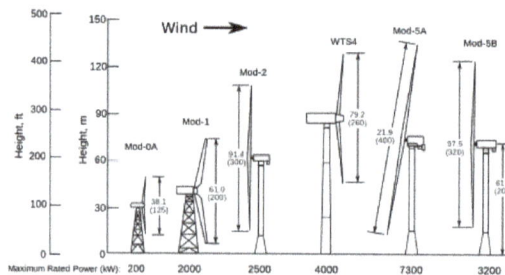

Comparison of NASA wind turbines

1973–2000

US Development

From 1974 through the mid-1980s the United States government worked with industry to advance the technology and enable large commercial wind turbines. The NASA wind turbines were developed under a program to create a utility-scale wind turbine industry in the U.S. With funding from the National Science Foundation and later the United States Department of Energy (DOE), a total of 13 experimental wind turbines were put into operation, in four major wind turbine designs. This research and development program pioneered many of the multi-megawatt turbine technologies in use today, including: steel tube towers, variable-speed generators, composite blade materials, partial-span pitch control, as well as aerodynamic, structural, and acoustic engineering design capabilities. The large wind turbines developed under this effort set several world records for diameter and power output. The MOD-2 wind turbine cluster of three turbines produced 7.5 megawatts of power in 1981. In 1987, the MOD-5B was the largest single wind turbine operating in the world with a rotor diameter of nearly 100 meters and a rated power of 3.2 megawatts. It demonstrated an availability of 95 percent, an unparalleled level for a new first-unit wind turbine. The MOD-5B had the first large-scale variable speed drive train and a sectioned, two-blade rotor that enabled easy transport of the blades. The 4 megawatt WTS-4 held the world record for power output for over 20 years. Although the later units were sold commercially, none of these two-bladed machines were ever put into mass production. When oil prices declined by a factor of three from 1980 through the early 1990s, many turbine manufacturers, both large and small, left the business. The commercial sales of the NASA/Boeing Mod-5B, for example, came to an end in 1987 when Boeing Engineering and Construction announced they were "planning to leave the market because low oil prices are keeping windmills for electricity generation uneconomical."

Later, in the 1980s, California provided tax rebates for wind power. These rebates funded the first major use of wind power for utility electricity. These machines, gathered in large wind parks such as at Altamont Pass would be considered small and un-economic by modern wind power development standards.

Danish Development

A giant change took place in 1978 when the world's first multi-megawatt wind turbine was constructed. It pioneered many technologies used in modern wind turbines and allowed Vestas, Siemens and others to get the parts they needed. Especially important was the novel wing construction using help from German aeronautics specialists. The power plant was capable of delivering 2MW, had a tubular tower, pitch controlled wings and three blades. It was built by the teachers and students of the Tvind school. Before completion these "amateurs" were much ridiculed. The turbine still runs today and looks almost identical to the newest most modern mills.

Danish commercial wind power development stressed incremental improvements in capacity and efficiency based on extensive serial production of turbines, in contrast with development models requiring extensive steps in unit size based primarily on theoretical extrapolation. A practical consequence is that all commercial wind turbines resemble the *Danish model*, a light-weight three-blade upwind design.

All major horizontal axis turbines today rotate the same way (clockwise) to present a coherent view. However, early turbines rotated counter-clockwise like the old windmills, but a shift occurred from 1978 and on. The individualist-minded blade supplier Økær made the decision to change direction in order to be distinguished from the collective Tvind and their small wind turbines. Some of the blade customers were companies that later evolved into Vestas, Siemens, Enercon and Nordex. Public demand required that all turbines rotate the same way, and the success of these companies made clockwise the new standard.

Self-sufficiency and back-to-the-land

In the 1970s many people began to desire a self-sufficient life-style. Solar cells were too expensive for small-scale electrical generation, so some turned to windmills. At first they built ad-hoc designs using wood and automobile parts. Most people discovered that a reliable wind generator is a moderately complex engineering project, well beyond the ability of most amateurs. Some began to search for and rebuild farm wind generators from the 1930s, of which Jacobs Wind Electric Company machines were especially sought after. Hundreds of Jacobs machines were reconditioned and sold during the 1970s.

Following experience with reconditioned 1930s wind turbines, a new generation of American manufacturers started building and selling small wind turbines not only for battery-charging but also for interconnection to electricity networks. An early example would be Enertech Corporation of Norwich, Vermont, which began building 1.8 kW models in the early 1980s.

In the 1990s, as aesthetics and durability became more important, turbines were placed atop tubular steel or reinforced concrete towers. Small generators are connected to the tower on the ground, then the tower is raised into position. Larger generators are hoisted into position atop the tower and there is a ladder or staircase inside the tower to allow technicians to reach and maintain the generator, while protected from the weather.

21st Century

Size comparison of modern wind turbines

As the 21st century began, fossil fuel was still relatively cheap, but rising concerns over energy security, global warming, and eventual fossil fuel depletion led to an expansion of interest in all available forms of renewable energy. The fledgling commercial wind power industry began expanding at a robust growth rate of about 25% per year, driven by the ready availability of large wind resources, and falling costs due to improved technology and wind farm management.

The steady run-up in oil prices after 2003 led to increasing fears that peak oil was imminent, further increasing interest in commercial wind power. Even though wind power generates electricity rather than liquid fuels, and thus is not an immediate substitute for petroleum in most applications (especially transport), fears over petroleum shortages only added to the urgency to expand wind power. Earlier oil crises had already caused many utility and industrial users of petroleum to shift to coal or natural gas. Wind power showed potential for replacing natural gas in electricity generation on a cost basis.

Technological innovations continues to drive new developments in the application of wind power. By 2015, the largest wind turbine were 8MW capacity Vestas V164 for offshore use. By 2014, over 240,000 commercial-sized wind turbines were operating in the world, producing 4% of the world's electricity. Total installed capacity exceeded 336GW in 2014 with China, the U.S., Germany, Spain and Italy leading in installations.

Floating Wind Turbine Technology

Offshore wind power began to expand beyond fixed-bottom, shallow-water turbines beginning late in the first decade of the 2000s. The world's first operational deep-water *large-capacity* floating wind turbine, Hywind, became operational in the North Sea off Norway in late 2009 at a cost of some 400 million kroner (around US$62 million) to build and deploy.

These floating turbines are a very different construction technology—closer to floating oil rigs rather—than traditional fixed-bottom, shallow-water monopile foundations that are used in the other large offshore wind farms to date.

By late 2011, Japan announced plans to build a multiple-unit floating wind farm, with six 2-megawatt turbines, off the Fukushima coast of northeast Japan where the 2011 tsunami and nuclear disaster has created a scarcity of electric power. After the evaluation phase is complete in 2016, "Japan plans to build as many as 80 floating wind turbines off Fukushima by 2020" at a cost of some 10-20 billion Yen.

Airborne Turbines

Airborne wind energy systems use airfoils or turbines supported in the air by buoyancy or by aerodynamic lift. The purpose is to eliminate the expense of tower construction, and allow extraction of wind energy from steadier, faster, winds higher in the atmosphere. As yet no grid-scale plants have been constructed. Many design concepts have been demonstrated.

References

- Sathyajith, Mathew (2006). Wind Energy: Fundamentals, Resource Analysis and Economics. Springer Berlin Heidelberg. pp. 1–9. ISBN 978-3-540-30905-5.

- Lucas, Adam (2006). Wind, Water, Work: Ancient and Medieval Milling Technology. Brill Publishers. p. 105. ISBN 90-04-14649-0.

- Ahmad Y Hassan, Donald Routledge Hill (1986). Islamic Technology: An illustrated history, p. 54. Cambridge University Press. ISBN 0-521-42239-6.

- Lucas, Adam (2006). Wind, Water, Work: Ancient and Medieval Milling Technology. Brill Publishers. p. 65. ISBN 90-04-14649-0.

- Mark Kurlansky, Salt: a world history,Penguin Books, London 2002 ISBN 0-14-200161-9, pg. 419

- Meri, Josef W. (2005). Medieval Islamic Civilization: An Encyclopedia. 2. Routledge. p. 711. ISBN 0-415-96690-6.

- History of Wind Energy in Cutler J. Cleveland,(ed) Encyclopedia of Energy Vol.6, Elsevier, ISBN 978-1-60119-433-6, 2007, pp. 421-422

- Erich Hau, Wind turbines: fundamentals, technologies, application, economics, Birkhäuser, 2006 ISBN 3-540-24240-6, page 32, with a photo

- Alan Wyatt, Electric Power: Challenges and Choices,(1986),Book Press Ltd., Toronto, ISBN 0-920650-00-7 , page NN

- Paul Gipe Wind Energy Comes of Age, John Wiley and Sons, 1995 ISBN 0-471-10924-X, Chapter 3

- Clive, P. J. M., Windpower 2.0: technology rises to the challenge Environmental Research Web, 2008. Retrieved: 9 May 2014.

- Clive, P. J. M., The emergence of eolics, TEDx University of Strathclyde (2014). Retrieved 9 May 2014.

- The World Wind Energy Association (2014). 2014 Half-year Report. WWEA. pp. 1–8.

- Griffith, Saul. "High-altitude wind energy from kites! (video)". Retrieved 5 March 2014.

- Warnes, Kathy. "Poul la Cour Pioneered Wind Mill Power in Denmark". History, because it's there. Retrieved 20 January 2013.

- "BTM Forecasts 340-GW of Wind Energy by 2013". Renewableenergyworld.com. 27 March 2009. Retrieved 29 August 2010.

- "Japan Plans Floating Wind Power Plant". Breakbulk. 16 September 2011. Retrieved 12 October 2011.

- Yoko Kubota Japan plans floating wind power for Fukushima coast Reuters, 13 September 2011. Accessed: 19 September 2011.

- Grove-Nielsen, Erik. Økær Vind Energi 1977 - 1981 Winds of Change. Retrieved: 1 May 2010.

- Ramsey Cox (February–March 2010). "Water Power + Wind Power = Win!". Mother Earth News. Retrieved 3 May 2010.

Permissions

Index